ADVANCE PRAISE FOR

The Carbon Buster's Home Energy Handbook

This really is a great resource! For all of us who want to reduce
our environmental impact *and save money*, here is the invaluable,
practical resource we've been looking for. Godo Stoyke presents a clearly
explained outline of what to do, where to start, and what the costs
and benefits will be. If you are concerned about global warming,
rising fuel costs, or the world we will leave our grandkids,
read this book, and tell your friends.

— ALEXIS KAROLIDES, AIA, LEED AP, Principal,
Breakthrough Design, Rocky Mountain Institute

The Carbon Buster's Home Energy Handbook offers a
treasure-trove of practical information that will help homeowners
reduce both their energy consumption and their carbon footprint.
For those wanting to make a difference with their homes and
their lifestyles, this book is a great starting point.

— ALEX WILSON, author of *Your Green Home* and
Executive Editor, *Environmental Building News*

This book is a dream come true for anyone who wants to
reduce their carbon emissions in the most cost-effective way.
Godo Stoyke plays the carbon numbers like Mozart played the
piano — with joy, inspiration, and accuracy. It should be
adapted for every country in the world.

— GUY DAUNCEY, author of *Stormy Weather: 101 Solutions
to Global Climate Change*

This book offers real solutions for reducing our energy use and
helping our environment. The practical suggestions in *The Carbon
Buster's Home Energy Handbook* can help an individual reduce their
greenhouse gas emissions and save money on energy bills.

— DOUG ANDERSON, ENERGY STAR program,
US Environmental Protection Agency

At last, a book written in plain English that guides you step-by-step through the best choices to make for reducing both energy costs and carbon emissions at a family level. Godo Stoyke has crunched the efficiency numbers so you don't have to.... and offers a return on investment for household, vehicle, and lifestyle that beats the best financial instruments out there — all to the good of the planet.

— RICHARD FREUDENBERGER, publisher, *BackHome Magazine*

Godo Stoyke's *Carbon Buster's Home Energy Handbook* is a timely and practical how-to guide to show each of us how we can contribute to a more sustainable world by reducing our carbon footprint. As a home owner and an ecological economist specializing in full-cost accounting, I found in this book a plethora of wonderful tips for saving energy and money, and tools to weigh various energy efficiency options. I'm eager to get on with implementing many of Godo's ideas in my own home!

— MARK ANIELSKI, ecological economist, and author of *The Economics of Happiness*

The Carbon Buster's Home Energy Handbook brings into focus the environmental impacts of personal choices around energy use. Godo Stoyke brings us out of hiding and calls us to action; no longer can we say that the efforts of an individual are too small to matter.

— PAUL SCHECKEL, energy analyst, and author of *The Home Energy Diet*

THE
CARBON BUSTER'S
HOME ENERGY
HANDBOOK

SLOWING CLIMATE CHANGE AND SAVING MONEY

GODO STOYKE

NEW SOCIETY PUBLISHERS

Cataloging in Publication Data:
A catalog record for this publication is available from the National Library of Canada.

Cover design by Diane McIntosh. Images: iStock/Photodisc.

Printed in Canada. First printing January 2007.

New Society Publishers acknowledges the support of the Government of Canada through the Book Publishing Industry Development Program (BPIDP) for our publishing activities.

Paperback ISBN-13: 978-0-86571-569-1
Paperback ISBN-10: 0-86571-569-6

Inquiries regarding requests to reprint all or part of *The Carbon Buster's Home Energy Handbook* should be addressed to New Society Publishers at the address below.

Disclaimer
Seek professional help for installations in your home where required, and obey all local electrical, building and safety codes. The author of this book has used his best effort in preparing the information for this book. Though we believe the information in this book to be accurate, the author and the publisher make no warranty of any kind, expressed or implied, with regard to the performance of the products described in this book, or for any procedures described herein. The author and publisher shall not be liable in any event for incidental or consequential damages in connection with, or arising out of, the use of these products or procedures.

Carbon Busters® is a registered trademark of Carbon Busters Inc. Other trademarks or registered trademarks are the property of their respective owners.

Financial self-disclosure: The author of this book owns stock in Ballard Power Systems Inc., a manufacturer of hydrogen fuel cells.

To order directly from the publishers, please call toll-free (North America) 1-800-567-6772, or order online at newsociety.com

Any other inquiries can be directed by mail to:

New Society Publishers
P.O. Box 189, Gabriola Island, BC V0R 1X0, Canada   1-800-567-6772

New Society Publishers' mission is to publish books that contribute in fundamental ways to building an ecologically sustainable and just society, and to do so with the least possible impact on the environment, in a manner that models this vision. We are committed to doing this not just through education, but through action. We are acting on our commitment to the world's remaining ancient forests by phasing out our paper supply from ancient forests worldwide. This book is one step toward ending global deforestation and climate change. It is printed on acid-free paper that is **100% old growth forest-free** (100% post-consumer recycled), processed chlorine free and printed with vegetable-based, low-VOC inks. For further information, or to browse our full list of books and purchase securely, visit our website at: www.newsociety.com

NEW SOCIETY PUBLISHERS   www.newsociety.com

This book is dedicated with love to my sister,
Birte Livia Stoyke (1969-2002),
who taught me how to live life to the fullest,
and to my parents, Eckhart and Heidi Stoyke,
who taught me everything else.

To find out about the charitable activities of
the Livia Stoyke Foundation, go to Livia.ca.

Contents

Acknowledgments

I WOULD LIKE TO THANK Chris and Judith Plant and their staff at New Society Publishers for enthusiastically picking up on, and promoting, the concept of carbon busting. Also, for the work of their excellent editing and design staff, and for the breathtaking cover. I would particularly like to thank Sue Custance, production coordinator, Ingrid Witvoet, editor, and Diane McIntosh, cover designer. Special thanks also to Richard Freudenberger, technical editor, for providing extensive feedback and thoughtful commentary on all parts of the manuscript. Finally, thanks to Beth Anne Sobieszczyk and Sara Reeves, marketing specialists, for their efforts in promoting the book. It's been a pleasure working with this group of intelligent and perceptive professionals.

For the contents of the book, I would like to thank one person and one organization in particular for inspiration: one is my father, Eckhart Stoyke, energy efficiency mentor, and the second is the Rocky Mountain Institute.

Eckhart has always had a passion and a knack for energy efficiency, whether it was working for Shell in France, Lufthansa in Germany, Polychemical Industries in Alberta, or for Dow Chemical, at the Elbe River or in Iran. At Dow Chemical, in 1978 as project engineer for an ethylene liquefaction facility, he proposed propylene as the coolant instead of the then-prevalent Freon, an ozone destroying CFC, nine years ahead of the original Montreal Protocol. He also designed a rail car tank wash facility that, unlike the external bid, included a method of cleaning washwater of chlorinated hydrocarbons (the proposal also cut costs by over 50 percent). When he represented the Western European Chemical Manufacturers at the United Nations Maritime Organization IMCO in connection with maritime pollution prevention in 1980, he proposed a slop-reduction and chemical product recovery system for marine tankers that not only cost-effectively recovered 90 percent more residues for clients, but at the same time facilitated the adoption of Marpol 77, a convention to protect the marine environment from chemicals.

For 12 years, Eckhart then served as the sole energy consultant of the nearly 200 Edmonton Public Schools in Edmonton, Canada. Faced with virtually no budget, and the mandate to create an effective energy management program, he was forced to rely entirely on his wiles for implementation. Choosing a no-cost and low-cost approach to efficiency, Eckhart applied a soft, education- and people-based approach to efficiency, a challenge for a mechanical engineer — or any engineer for that matter.

By achieving no-cost savings of Cdn$150,000 in the first year and Cdn$500,000 in the second year, he convinced the school board to allocate 10 percent of utility funds for five years for retrofit technologies that achieved over Cdn$24 million in savings for the board over 11 years. Small wonder his program at the school board was recognized with the Best Corporate Program Award of the Department of Energy, Mines and Resources of the Canadian Government, and the Best Corporate Program Award of the Association of Energy Engineers in 1989. From there it was only a small step for him to start his own company, Carbon Busters Inc., and to quit his job in 1993 (at age 58), to offer similar programs to educational institutions, municipalities and other large facilities in the US, Canada and Europe.

A few months after the company was founded I was invited to join Eckhart's new venture. While I had been self-employed in the field of energy conservation and environmental education before joining Carbon Busters, the difference was that at Carbon Busters I actually got paid for my work. Now, $16 million in client savings and 120 million pounds in prevented greenhouse gas emissions later, we have never looked back.

The second entity to inspire me, and to inspire the staff here at Carbon Busters, is the Rocky Mountain Institute (rmi.org). From the mid-1970s writings on the Soft Energy Path, to the 1982 founding of the Rocky Mountain Institute by Hunter and Amory Lovins and beyond, we have followed the progress of Amory Lovins and the RMI folks breathlessly. Their unique combination of optimism, pragmatism, creative problem-solving and intellectual brilliance sets them in a class apart. Surely, the Rocky Mountain Institute (and, incidentally, the Grameen Foundation, Grameen.org, which helps provide microcredits for the poor, especially women) are among the brightest forces for good on the planet.

Next, I would like to thank the talented staff at Carbon Busters for their numerous hours of meticulous background research for this book, and for the hours spent in trying to come up with better book titles and cover design ideas. Carbon Busters Shanthu Mano, Anna Klimek, Pat Roth, Cecylia Krzesinski, Heidi Stoyke, and particularly Amy Mireault provided ample research and insights. Michèle Elsen and Claudia Bolli, in addition to their research, spent countless hours testing the latest electronic devices and compact fluorescent lights for effectiveness and energy efficiency, scouring Edmonton's environs to leave no stone unturned for opportunities to connect consumer products to a power meter and diligently recording the results. Richard Krause and Eckhart Stoyke provided additional research on numerous specific efficiency projects (Richard and Claudia also provided ideas for the cartoons). Asmus Stoyke provided his engineering skills, and spreadsheet brilliance, for projects ranging from calculations of U-values, R-values and heat transfer, to aspect ratios and other arcane mysteries of energy conservation and efficiency. Thanks to HVAC specialist Michael Beblo for information on cooling efficiency.

Visual designers Gabriel Wong (gabriel@designustrator.com) and Patricia Begley (neuroticgirl@hotmail.com) ably created most of the illustrations in the book, and Gabriel also provided the initial inspiration for the dollar leaf concept on the cover. Thanks also to visual designer Rina Chan (rchan_@hotmail.com) for photos of compact fluorescent lights and energy-efficient cars.

Thanks to the following for generously providing illustrations for this book: renowned Canadian architect Douglas Cardinal (djcarchitect.com), RMI (rmi.org), the ING Bank (ing.com), green builder Peter Amerongen of Habitat Studio & Workshop Ltd. (habitat-studio.com), green architect Jorg Ostrowski (ecobuildings.net), the Kuhn Rikon Corporation (kuhnrikon.com), Volkswagen AG (volkswagen-ag.de), and Adam Opel GmbH (opel.de), and to Johannes Wheeldon (biodieselsolutions.ca) and Joey Hundert for providing pricing information for biodiesel. Thanks to Shawn Murphy (smurp.com) for providing information on Linux sleep mode (reproduced more or less verbatim). Thanks to Roger Huber at swisssolartech.com for providing detailed cost and benefit analyses for solar hot water collector setups for Los Angeles, Las Vegas, and Edmonton.

Special thanks to the vendors and individuals who provided us with access to their equipment for energy efficiency measurements:

Westworld Computers: westworld.ca, especially Calvin Anderson and Robyn Shilling.

Generation Electronics: generation.ab.ca, especially Mike Jackman and Ted Zanetic.

The Brick: thebrick.com, especially Gregg Wolkowski.

Three Hat: threehat.com, especially Greg MacIntyre and Scott Fisher.

Programmer Shawn Murphy: smurp.com

Carbon Busters staff: carbonbusters.org

I would also like to thank the following for reviewing parts or all of the draft manuscript for the book and providing numerous suggestions and improvements: Mark Anielski (anielski.com), Ted Wolff-von Selzam, Shawn Murphy (smurp.com), Brian Johnston, Melvin Neufeld, Michael Beblo, Eckhart Stoyke, Michèle Elsen, Claudia Bolli, Richard Krause, Asmus Stoyke, and Shanthu Mano.

Thanks to the friendly staff at the St. Albert library, for providing one of the best selections of books for a library of this size I have ever seen.

Thanks to our far-flung family friends, the Wolff and Wolff-von Selzam clans, for providing electronic long-distance encouragement and cheering via iChat from Ontario, Florida and Germany.

I would like to especially thank my lovely wife, Shanthu Mano, for providing numerous suggestions for improving the approach and direction of the Carbon Buster's Handbook, for being a tireless advocate of our principles (and myself), and lastly, for shouldering part of my house work during crunch times. Finally, I would like to thank my 6-year old son, Calan, for being unfailingly understanding when, entering the writing home-stretch, I was on occasion not available for a light saber tournament, not even a quick one!

Congratulations!

BY PICKING UP THIS BOOK, you have shown that you care about our environment and that you are ready to be part of the solution.

Whether you buy just one compact fluorescent light bulb or retrofit your whole house, your contribution is vitally important to slow the environmental tailspin.

If we achieved the energy efficiency of Japan, we could save US$470 billion per year, or $1,400 for each American and Canadian.[1,2,3] And Japan is inefficient compared to the energy-efficiency technology available to you in 2006.

Throughout the book, you will notice that costs, distances, weights and other measures are referenced. Unless otherwise noted, these are presented in standard US measure (see "Conventions" heading in Chapter 2). There is also a metric conversion table at the end of the book (Appendix K) for most measurement conversions.

Fortune favors the brave.
— Virgil, Aeneid

Thanks for making the world a better place to live in!

Carbon Busting for Fun and Profit

THIS BOOK IS BASED ON A VERY SIMPLE PREMISE: environmental protection is cheap!

If you add up all the services provided by a healthy biosphere, and compare these to the costs associated with living an energy-, material- and pollution-intensive life-style, the environmentalist approach comes out far ahead every time.

Let's take a simple example: the Kyoto Protocol.

The Kyoto Protocol became a legally binding treaty on February 16, 2005.[4] The protocol aims to reduce the world's anthropogenic (human-induced) greenhouse gas emissions (carbon dioxide, methane and a few other gases). In fact, the goal of this international agreement is to reduce greenhouse gas emissions of participating industrialized countries by five percent or more below 1990 levels by 2008 to 2012.[5]

> **Quick Fact**
>
> **carbon busting** (Kär'bən bus'ting) *v. Informal.* employing the most cost-effective methods to drastically lower greenhouse gas emissions.

The treaty has been ratified by 55 nations, including all the major industrialized nations, with the notable exceptions of Australia and the US. Kyoto was signed under Bill Clinton, but not ratified under George W. Bush, despite the fact that the US is the world's largest single greenhouse gas emitter.

The Kyoto Protocol, although perhaps the most significant global agreement ever to come into force, is actually only a first baby step. Based on the world's most comprehensive scientific analysis, as represented by the findings of the Intergovernmental Panel on Climate Change (IPCC), we have to reduce our greenhouse gas emissions by 60 to 80 percent below 1990 levels just to stabilize today's already elevated atmospheric greenhouse gas levels.

Some individuals and organizations have argued that Kyoto would be too costly to implement, claiming costs as high as $359 billion for the US alone, based on costs of $67 to $348 per tonne (Note: one metric tonne — international symbol "t" — is equal to 2,204 pounds, or 1.1 US tons.). [6]

However, careful analysis shows that implementing the Kyoto protocol is not only affordable, but highly profitable. As Amory Lovins, head of the renowned efficiency think tank Rocky Mountain Institute points out: "The amount is about right. Only the sign has to be reversed!"[7,8,9]

For example, while some oil companies, notably among them American-based ExxonMobil (with its chain of Exxon, Esso and Mobil gas stations), were still lobbying heavily against the Kyoto protocol on the grounds that it was "too expensive," manufacturer 3M and European-based oil companies BP and Shell announced that they had already exceeded the requirements of the Kyoto Protocol, at no cost, or at huge profits, due to energy savings.[10]

For example:

- Between 1975 and 1999, 3M saved $827 million and achieved energy efficiency improvements of 58 percent per unit of production.[11]
- Shell reduced its emissions 10 percent below 1990 levels (exceeding Kyoto targets by 100 percent), at no net economic cost, and achieved this in 2002, six years ahead of the 2008 to 2012 target.[12]
- BP achieved savings of $650 million from emissions reductions.
- Interface, the largest flooring manufacturer in the world and a pioneer in offering service models for carpeting, saved $185 million. (The company rents out long-lasting quality carpets.)[13]

Apparently, ExxonMobil has spent $12 million since 1998 in an attempt to convince politicians not to take action on climate change.[14] It makes one wonder by how much ExxonMobil could have exceeded the Kyoto Protocol requirements, and how much money ExxonMobil *could* have saved for its shareholders, if it had invested those $12 million in energy-efficient operations instead.

Quick Fact

Preventing climate change is hugely profitable, not costly.

Let's take another example: Carbon Busters Inc., the company from which the title of this book is derived, and which supplies the author with his monthly income.

Carbon Busters advises large facility operators, for example school boards and municipalities, on how to run their buildings more energy efficiently. With only a handful of staff, Carbon Busters helped its clients reduce greenhouse gas emissions by a stunning 121 million pounds of carbon dioxide.

What was the cost of creating these savings?

Well, here is the interesting part: The building operators didn't lay out a single cent to create these enormous greenhouse gas savings.

Instead, the whole program was financed by a fraction of the achieved utility savings. The rest of the money went to buy school books and basketballs, to reduce school taxes, and to purchase efficiency technologies to create even higher savings.

So, the "cost" of creating 121 million pounds of greenhouse gas reductions actually turned out to be $16 million in utility savings.

Remember the projected cost of saving one tonne of carbon? $67 to $348. Well, it turns out that the large building operators supported by Carbon Busters were actually *saving* $318 for each metric tonne of carbon dioxide emissions prevented, or $1,163 per tonne of carbon.

And before you think that these kinds of savings are only possible in wasteful North America, it pays to keep in mind that a lot of these Carbon Busters savings were created in supposedly energy-efficient Europe. And, understandably, even higher savings can be achieved in many developing countries that often rely on even more ancient and less efficient technology than that in use in the average American household or Canadian factory.

Which leads us to the purpose of this book: carbon busting for fun and profit, and how you personally (and the environment generally) can benefit from the energy efficiency revolution.

> ## Quick Fact
>
> The US and Canada alone can realize over $300 billion in annual, cost-effective energy savings.

This book describes step-by-step where the modern carbon miser can do the most good, and where your bank account can grow the fastest.

By following the efficiency measures outlined here, a typical family can save $17,000 in energy costs over five years, and exceed the requirements of the Kyoto Protocol in the home arena by 860 percent, with an average payback of 3.5 years.

It is left to you, the reader, to decide if you want to maximize your dollar return by choosing the path of the Carbon Miser (*Carbon Miser*, p. 15), or maximize your carbon savings by following the road of the Carbon Buster (*Carbon Buster*, p. 15), for that extra warm feeling that comes from going beyond the call of common duty.

Either way, I hope that you will enjoy the journey, give an environmental boost to our planet, and save a bundlle.

2

Conventions and Assumptions

THE PURPOSE OF THIS BOOK is to show you the most effective, and most cost-effective, ways to reduce your carbon emissions, and the best ways to reduce your family's energy bills.

While there are many lifestyle changes and behaviors you can adopt to reduce your ecological footprint, this book primarily describes products and technologies.

What these technologies demonstrate is that you can solve the greenhouse gas problem single-handedly, or at least your family's contribution to it, simply by making good buying choices.

Furthermore, doing so will require no great sacrifices on your part. In fact, many of the environmentally superior products are also of higher quality, more comfortable, healthier for you, and more fun to use.

Finally, in addition to their environmental benefits, these products will save you a substantial amount of your annual expenses, allowing you to redirect the savings to enjoyable (and hopefully environmentally benign) purposes, or simply allowing you to work less and spend more meaningful time with your family.

If you are willing to go beyond measures offered in this book through behavior changes (for example, turning off the lights in unoccupied rooms), please be assured that I strongly encourage you to do so — you are benefiting the environment even more, and those energy savings cost nothing and therefore have an immediate financial payback. You can probably reduce your carbon emissions by an additional 10 or 20 percent.

Yet even without that, a typical family can cut its annual carbon emissions by over 70

> ## Quick Fact
>
> Many of the environmentally superior products are also of higher quality, more comfortable, healthier for you, and more fun to use.

percent, and its energy bills by over $3,500 per year, based on the technology presented in this book.

Assumptions and Methodology

If we look at the average US family, you will find 2.59 persons per household (3.14 per family), and 1.7 cars, all using energy in the form of electricity, natural gas, propane, oil, kerosene *and* wood (see Chapter 4, "Your Family's Carbon Pie," for details).[15, 16]

Obviously, no family with these exact demographics exists.

Therefore, to allow us to establish a base-case scenario against which we can measure savings, we are using the consumption of a typical American family. This household has three or four members, two cars, and a free-standing house of about 2,000 to 2,100 square feet. The energy consumption for our fictitious family is largely based on energy consumption and carbon emissions data presented in a 2005 *Discover* magazine article by Richard Coniff (adjustments have been made for CO2 conversion factors used in this book).[17] These numbers in turn are based on statistics compiled by Lawrence Berkeley National Laboratory, the 2001 US Department of Energy Residential Consumption Survey, the 2001 National Household Travel Survey and Oak Ridge National Laboratory.

Also, to calculate paybacks, there is often a big difference in net savings depending on whether you are only paying for the difference between a more or less efficient device (incremental savings), or whether you have to pay the full cost to replace a device in good working order.

These are the assumptions with respect to your family's appliances and equipment: You need a new washer/dryer combo, and you are in the market for new cars. Your refrigerator is from 1993, and works fine. You are using high-efficiency (low-flow) showerheads. You have a medium-efficiency gas furnace that is not giving you any trouble, and you don't intend to replace your computer.

Correcting for Savings Interactions

Many heating and cooling energy measures are additive: for example, if you buy a better window, and insulate your attic, savings from these measures can be added up. However, some of these measures will partially reduce the savings potential of others. For example, if you install a 95 percent condensing furnace instead of an 80 percent medium-efficiency furnace, your savings potential of all other heating efficiency measures has just gone down by 15.8 percent.

These savings interactions have generally been considered in the "Carbon Buster" and "Carbon Miser" scenarios described on pages 15 to 18. Also, alternate calculations are often provided in the appendix E (page 140), dependent on which measures you implement, so that you can assemble your own list of measures, and still have reasonably accurate savings estimates.

Conventions
Currency Used
Unless specifically noted otherwise, all currencies are expressed in 2005/2006 US dollars. To correct for inflation in later years, you can add about three percent per year. Inflation applies to both technology and energy prices, which largely cancel each other out. However, the January 2006 inflation spike of 3.99 percent was largely due to increasing energy prices. If this trend continues, the cost of efficiency technology will increase more slowly than energy prices, giving you better paybacks and returns.

Rounding
Results for carbon savings are usually rounded to the closest 100 pounds. Precise, unrounded numbers are provided in the appendix.

EPA Mileage Ratings
For composite miles per gallon figures, 55 percent city and 45 percent highway driving is assumed. For calculations, an additional 15 percent is deducted from EPA fuel consumption ratings to approximate real-life efficiency.

Energy Prices
All calculations are based on current US national average energy prices at the time of writing in 2005 and 2006 (see Appendix G, p. 145).

Life-cycle Environmental Impact
This study looks primarily at direct (Scope 1) emissions caused by energy consumption. It also includes indirect home emissions created at the power plant due to combustion for the production of electricity (Scope 2 indirect emissions). It does not include upstream emissions caused in the production of fuels (Scope 3 indirect emissions). Scope 3 emissions include energy required for the exploration, extraction, processing and transportation of fuels. These emissions vary, but may add something like 10 percent to the emissions of any given unit of fuel.

Also, this book focuses on carbon dioxide, which is generally considered to be the most important greenhouse gas. However, energy savings generally also lead to reductions in other greenhouse gases such as nitrous oxides or methane, or to reductions in particulate emissions.

Finally, with a few exceptions, this study does not examine the life-cycle emissions of fuels or devices. Life-cycle emissions may include the emissions caused by the construction of facilities, e.g.: oil rigs, power plants or wind turbines. It can also include the energy required to build devices. With energy-consuming devices, the energy required for manufacture is often far less, by a factor of ten or so, than the energy consumed over the life of the device (for example,

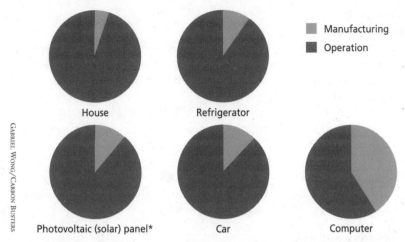

GABRIEL WONG/CARBON BUSTERS

2.1: *Energy costs of houses, cars, computers, refrigerators and solar panels: manufacturing vs. operating costs.[18] * Energy production vs. energy consumption during manufacturing for solar panels.*

cars, houses or fridges, Figure 2.1). Notable exceptions are high-tech devices, such as computers. Even though computers use a fair amount of energy during normal usage, they contain highly refined components that require relatively high amounts of materials and energy to produce. Desktops require nearly as much energy to produce as they consume over their life, and notebooks may actually require more during manufacture.

Non-energy consuming products used in the home also require energy to manufacture. For example, according to a study commissioned by Marks & Spencer, the life-cycle energy requirements for a pair of blue men's polyester pants was just under 200 kWh (or just under $20, at current average electricity prices). Of this energy, 25 percent was due to manufacture, and 75 percent to subsequent washing and ironing.[19]

In the future, such life-cycle studies may become more important as we learn how to reduce our environmental impacts and our global ecological footprint.

Sources of Information
Sales Staff

There are many hardworking, well informed sales people that can advise you in the process of purchasing the best equipment. However, there are also a good number of sales people who will give poor advice due to ignorance, or in an effort to increase sales or earn a higher commission.

This is why it can pay off to deal with someone whom you trust, even if it means paying a few dollars more on a specific product. A trustworthy salesperson is worth his or her weight in gold. The good advice may pay for itself several times over if it best meets your needs. If in doubt, it is generally a good idea to rely on the advice of efficiency experts, government sources and consumer magazines instead of unknown sales staff who may stand to gain by providing suboptimal solutions to you.

Just two recent examples from my own circle of acquaintances: a car dealer discouraged a friend from buying a Toyota Prius hybrid car because, he claimed, they have "trouble starting in winter" (not true). It was more likely due to the lower commission potential of the fixed-price Prius.

Another dealer told a friend *not* to buy a high-efficiency furnace which was recommended during an energy audit, because "they are not worth it" (the dealer did not have any in stock).

Internet Information

The Internet is a fantastic source of information and becoming, if anything, more important as a research tool. However, it also removes editorial processes that can improve the quality of information in printed media. It is therefore doubly important to check sources on the Internet. Generally, the most reliable information can be found on sites from research institutions, governments and public interest non-profit organizations. With many other sites, you clicks your links and you takes your chances.

Interpreting Savings Figures

Savings data for particularly important savings measures discussed in the book are generally presented in a standard format (see below). Also, there are many more savings data listed for measures in Appendix G. What do the savings sections refer to?

Carbon Buster/Miser Recommendation: These point out measures and technologies that are among the most effective in reducing greenhouse gas emissions, and among the most profitable. They also form the basis of our cumulative savings results at the bottom of the page, starting at p. 43, and the summaries shown in Chapter 3: "For Busy People: Top Recommendations at a Glance." Carbon Buster recommendations are represented by a "leaf-in-a-light bulb" logo; Carbon Miser recommendations by Scrooge's top hat.

5-year savings: Savings in dollars, pounds (and metric tonnes) carbon dioxide, and in consumption units over five years are shown here. The carbon dioxide savings are generally rounded to the nearest 100 pounds.

Life-time savings: This section shows the savings over the life of the product. For example, if a washing machine on average lasts for eleven years, these numbers reflect annual savings times 11. Calculations for some items with a particularly long life have been artificially limited to less than the rated life-span, as you may not be that interested in power savings from LED Christmas lights 277 years from now! Conversely, in the rare cases where an item has a rated life of less than five years, we have multiplied the item's cost with the number of times you would have to purchase or install it to achieve at least those five years of service.

Incremental cost: This is the net difference between an efficient product and a standard product. It assumes that you are buying the product at any rate, and have to justify the savings only on the basis of the price difference between the two. In some cases, this incremental difference is easy to determine. For example, Sears offers both electric and gas dryer versions for many of its otherwise identical models. Volkswagen offers both gasoline and diesel versions of the Jetta.

In other cases it is not quite so simple. For example, there is no gasoline-only equivalent to the Toyota Prius hybrid car. In that case, we have used generally accepted estimates of the incremental costs, in this case, of paying for the hybrid engine.

In a few cases, the comparison is even more difficult. Take for example the two-seater, aluminum-framed Honda Insight hybrid car. Do we compare it to a five-seater Honda Civic? A two-seater sports car?

Many people have a difficult time comparing two cars differing only in styling, body color and number of cup holders, never mind radical departures in basic car design.

In those cases we have tried to decide on the basis of what most people would likely consider reasonable trade-offs.

New cost: These costs are assuming that the purchase has to justify itself entirely on the basis of energy and environmental savings. An example would be replacing a fridge that is working fine with a more energy efficient model. Costs generally include materials and labor. If the job would be done by most people themselves, we have also included a modest fee to yourself of $10/hour, to keep the calculations fair. It's not very much, but the work (generally) is not very hard, you can set your own hours, and, best of all, it's post-tax money! (Equivalent to $14 per hour in pre-tax income at a 30 percent tax rate.) Or, hire your niece Jane to do it, if she will work for $10 per hour.

Payback incremental / Payback new. IRR, CROI: paybacks are expressed in years, IRR in percentage points and CROI in pounds of carbon dioxide emissions prevented per dollar invested. The payback in years is a simple payback, i.e. the cost of the measure divided by annual savings, not accounting for the (interest) cost of money. Paybacks of zero to five years are generally considered excellent, five to ten years pretty good to reasonably good, and paybacks of 10 to 15 years or more are fair to poor. The lower the number of years, the better the payback. "Incremental payback" is based on incremental cost, "New payback" is based on new cost.

The IRR is the internal rate of return. It is calculated iteratively by taking the lifetime savings of a measure, subtracting the cost of implementation, and calculating the percent return per year of the remaining money over the life of the project. It is the equivalent interest rate earned over the lifetime of an investment. As such, it is directly comparable to other investment opportunities, such as, for example, government bonds yielding 5 percent, or mutual funds yielding 10 percent. The higher the percentage yield, the better the rate of return.

The carbon return on investment (CROI) in pounds per dollar, indicates how effective each dollar invested is in preventing the emission of additional carbon dioxide into the atmosphere. The higher the number, the better.

The payback calculations in this book are based entirely on personal financial gain, and reductions in greenhouse gas emissions (mostly carbon dioxide). Yet, that does not have to be *your* sole criterion.

For some people, knowing that they have prevented some asthma deaths, reduced the decay of historic monuments from acid rain, or slowed negative climate changes, may be payment enough. After all, many individuals spend all kinds of money on products without ever worrying about their payback. What is the payback on a $100,000 Mercedes, $3,000 leather car upholstery or a Ferrari 208?

If you feel good about a car that puts 60 percent less climate-damaging pollutants into the air, or a solar panel system that increases national (and personal) energy security, who is to say that alone isn't reason enough for spending some extra money?

Or, as someone said, "If we look at paybacks only, it could turn out that we cannot afford to keep the earth!"

3

For Busy People:
Top Recommendations at a Glance

From Carbon Miser to Carbon Buster

Carbon Miser

ONE APPROACH TO USING THIS BOOK is simply to look at your internal rate of return (IRR) for each of the proposed measures.

The IRR represents the net yield in percent per year, adjusted for the labor cost of your time, that you can expect for your investment. It allows you to directly compare the financial benefit of your investment to returns from, say, a mutual fund or government bonds.

You will find that many of the ideas in this book will give you a money return that would be suspected of being illegal, immoral or fattening by most financial advisors and mutual fund dealers in the world of investing.

So, this then is the path of the Carbon Miser: simply maximize your dollar return by choosing the investment with the highest return first, regardless of the environmental benefit.

This way, you pick the economic no-brainers. However, rest assured: every dollar saved also represents a huge benefit to the environment as well, since all savings translate into pounds and tonnes of carbon dioxide, sulphur dioxide, nitrous oxides and numerous other detrimental chemical emissions prevented.

Using the Carbon Miser's approach, and based on a typical family, you can save $14,120 in energy costs over the next five years ($2,824 annually), and reduce your annual carbon emissions by 40 percent. The internal rate of return for all measures combined is 62%, an excellent rate of return (Figure 3.1).

Carbon Buster

This approach emphasizes the benefit to the environment: what is the most effective way to reduce your carbon emissions?

Cost is still a factor, but the primary goal is to reduce your total annual carbon load.

While there are a few approaches that result in little or no financial gain (notably the important adoption of green power, outlined on page 95), the Carbon Buster will still realize substantial savings. In fact, the gross savings over five years will be higher by about $3,700, though the payback period is increased from 3.5 years to 5.4 years.

Measure	5-year Dollar Savings ($)	Carbon Pie Savings (%)	Cumulative $	Cumulative %
I: Toyota Echo (manual) as primary family car (5 seats), p. 47	5,037	12.4%	5,037.8	12.4%
I: Mercedes Smart Diesel car as secondary car (2 seats), p. 45	3,475	8.5%	8,512.0	20.9%
N: Replace 20 of your 25 lights with Compact Fluorescent Lights (CFLs), p. 56	350	1.6%	8,861.6	22.5%
N: Eliminate 90% of your power vampires, p. 67	524	2.4%	9,386.0	24.9%
I: Replace electric stove with natural gas stove, p. 79	209	1.4%	9,595.0	26.3%
I: Replace top-loading washer with front-loading model (water is heated by gas), p. 82	334	1.2%	9,928.8	27.4%
N: Use warm/cold instead of hot/hot wash mode (gas heated front-loading washer) p. 82	74	0.2%	10,003.0	27.6%
I: Replace electric dryer with gas dryer (washer is front-loader), p. 84	208	1.4%	10,211.1	29.1%
N: Put computer in sleep-mode, turn off when unused, p. 85	343	1.5%	10,554.3	30.6%
N: Replace your Christmas lights with LEDs, p. 95	106	0.5%	10,669.9	31.1%
N: Seal air leaks in your house, p. 107	855	2.3%	11,515.2	33.4%
N: Add R-40 cellulose insulation to unheated attic, p. 108	438	1.2%	11,952.7	34.6%
N: Install window kits (shrink foil) on 50% of your windows, p. 110	428	1.2%	12,380.4	35.8%
N: Get a tune-up for your furnace, p. 112	385	1.0%	12,765.3	36.8%
N: Add manual chimney cap to fireplace flue, p. 112	535	1.5%	13,309.9	38.3%
N: Buy electric-ignition tankless water heater, p. 115	614	1.7%	13,913.7	39.9%
N: Buy two low-flush toilets to replace 3.5 gallon toilets, p. 121	207	0.2%	14,120.3	40.2%

Total savings: 40% of your family's carbon pie, $14,120 saved over 5 years. Internal rate of return over the life of the measures: 62%.
I = incremental scenario (choose more efficient alternative when replacing worn-out model).
N = new scenario (old unit still functional, replaced with more efficient model).

3.1: *Carbon Miser recommendations: The top methods to save money on your energy bills.*

Using the carbon buster approach, you can cut your carbon emissions by an amazing 73 percent, and your energy costs by $17,805 over five years, or $3,561 per year — 57 percent. (Fig. 3.2).

Measure	5-year Dollar Savings ($)	Carbon Pie Savings (%)	Cumulative $	Cumulative %
I: Toyota Prius hybrid car as primary family car (5 seats), p. 42	6,722	16.5%	6,722.5	16.5%
I: Mercedes Smart Diesel car as secondary car (2 seats), p. 45	3,475	8.5%	10,197.7	25.1%
N: Use 50% biodiesel for Smart or other diesel, p. 49	0	7.5%	10,197.7	32.6%
N: Replace 20 of your 25 lights with Compact Fluorescent Lights (CFLs), p. 56	350	1.6%	10,547.3	34.1%
N: Install a solar tube in a frequently used area, p. 65	108	0.5%	10,655.6	34.6%
N: Eliminate 90% of your power vampires, p. 67	524	2.4%	11,179.9	37.0%
I: Replace electric stove with natural gas stove, p. 79	209	1.4%	11,389.0	38.4%
N: Purchase efficient cookware p. 80	137	0.4%	11,525.7	38.8%
I: Replace top-loading washer with front-loading model (water is heated by gas), p. 82	334	1.2%	11,859.5	39.9%
N: Use warm/cold instead of hot/hot wash mode (gas heated front-loading washer) p. 82	74	0.2%	11,933.6	40.1%
I: Replace electric dryer with gas dryer (washer is front-loader), p. 84	208	1.4%	12,141.8	41.6%
N: Put computer in sleep-mode, turn off when unused, p. 85	343	1.5%	12,484.9	43.1%
N: Replace your Christmas lights with LEDs, p. 95	106	0.5%	12,590.5	43.6%
N: Subscribe to green power, p. 95	0	15.2%	12,590.5	58.8%
N: Seal air leaks in your house, p. 107	855	2.3%	13,445.8	61.1%
N: Add R-40 cellulose insulation to unheated attic, p. 108	438	1.2%	13,883.3	62.3%
N: Add R-12 insulation to basement walls, p. 108	367	1.0%	14,249.9	63.3%
N: Install window kits (shrink foil) on 50% of your windows, p. 110	428	1.2%	14,677.6	64.5%
N: Get a tune-up for your furnace, p. 112	385	1.0%	15,062.5	65.5%

3.2: *Carbon Buster recommendations: The top methods to reduce your greenhouse gas emissions.*

Measure	5-year Dollar Savings ($)	Carbon Pie Savings (%)	Cumulative $	Cumulative %
N: Add motorized combustion air dampers to fresh-air intake for furnace, p. 112	171	0.5%	15,233.6	66.0%
N: Add manual chimney cap to fireplace flue, p. 112	535	1.5%	15,768.2	67.4%
N: Install a thermal solar collector for your hot water needs, p. 115	1,216	3.3%	16,984.6	70.7%
N: Buy electric-ignition tankless water heater, p. 115	614	1.7%	17,598.4	72.4%
N: Buy two low-flush toilets to replace 3.5 gallon toilets, p. 121	207	0.2%	17,805.0	72.6%

Total savings: 73% of your family's carbon pie, $17,805 saved over 5 years.
Internal rate of return over the life of the measures: 32%.
I = incremental scenario (choose more efficient alternative when replacing worn-out model).
N = new scenario.

4

Your Family's Carbon Pie

Wfor it?HAT IS YOUR ANNUAL OUTPUT OF CARBON, and how much will you pay

The latest national government averages of residential energy consumption for the US are from 2001. Based on these numbers, the average home energy use at that time led to the release of 36,000 pounds of carbon dioxide per year (Fig. 4.1).[20]

The average residential energy costs per household were $1,450, though costs were generally a lot higher in higher income families and larger homes. Neither of these figures include the carbon emissions and dollar costs of gasoline fuel for vehicles.

These figures are based on national averages, including a mix of fuels for home heating. Of course, actual homes don't typically use, for example, natural gas *as well as* fuel oil and propane for heating, just as there are very few homes (none, actually) that have the national average of 2.59 residents.

So, rather than look at a hypothetical average, we will look at a typical home. This will provide a useful basis for many homes and, with some adjustments for different equipment and fuel types, be helpful for the largest number of readers.

Interpreting Home Consumption Graphs

Mark Twain once said: "There are three kinds of lies: lies, damn lies, and statistics!"

While numbers are extremely important in interpreting the world around us, it is always important to check the source of the numbers, and how they

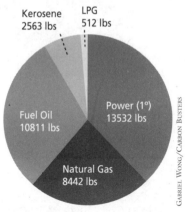

4.1: *2001 Average residential carbon dioxide emissions in the US per household, excluding car emissions, were 36,000 pounds. LPG = liquified petroleum gas. (Source: US Dept of Energy.)*

19

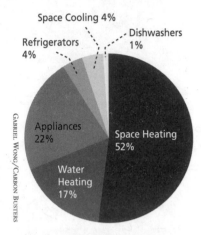

4.2: *Onsite Energy Use in Homes. (US Department of Energy).*[21]

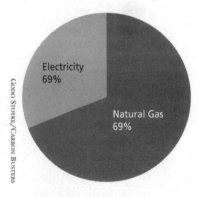

4.3: *The misleading presentation of onsite energy use: electrical power vs. natural gas, excluding primary energy use for electrical power.*[22]

Quick Fact

Beware the hidden cost of electricity: each kWh of power requires 3 kWh of primary energy to produce it (national average).

are used. This also applies to graphs on home energy use.

Take a look at a typical energy consumption graph for homes, this one from the US Department of Energy (Figure 4.2). At first glance it is quite innocuous, and similar to graphs used in numerous publications and utility company web-sites. It shows the onsite energy use of US homes. Graphs like these are not intended to be misleading. However, because of the hidden energy cost of electricity, they distort the primary energy picture.

You may notice that space heating (52 percent of onsite energy use) and water heating (17 percent) together make up 69 percent of energy use — an impressive amount of energy. In natural gas heated homes, the space heating and water heating would typically *both* be supplied by natural gas, while all the other uses in Fig. 4.2 would be supplied by electricity. So, if we add up all natural gas and power uses, respectively, we end up with a graph like Fig. 4.3. A full 69 percent of the energy used is supplied by natural gas, only 31 percent by power. It would be reasonable to conclude from this graph that most of the CO_2 emissions and most of the monetary costs for energy arise from the natural gas consumption, but this is not the case.

Electricity: The Most Expensive Form of Energy

Electrical power is not only the most versatile form of energy, it is also one of the hardest to create. In fact, for every kWh of electricity we use in our homes, typically 3 kWh or more of fossil fuels are burned at a power plant.

This means that for every one kWh of fossil-fuel derived power we use in our home, two kWh are lost at the plant as waste heat, and further losses are sustained through power transmission via high-voltage transmission lines and transformation of voltages. In this context, electricity is referred to as secondary energy, while the fossil fuels used to create the electricity are called primary energy.

Often, you do not have a choice with respect to the type of energy used. For example, if you want to use your computer, well, it has to be electricity. (Tiny methanol fuel cells may change this: they convert a primary fuel directly to electricity and may become more important in the very near future, especially for portable equipment like laptops and cell phones.)

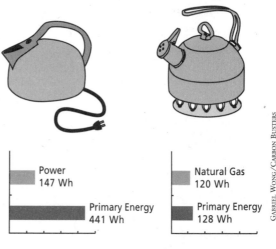

However, if you are using energy for heating purposes, using primary energy instead of secondary energy (electricity) is almost always much more efficient. Let's take the example of heating a kettle of water (Fig. 4.4).

4.4: *Primary and secondary energy consumption of a water kettle to heat 0.5 L (about one pint) of water, based on heat source (after Seifried).*[23]

If you use electricity to heat the water, you will consume 441 Watt-hours of energy. However, if you use natural gas to heat the water directly, you only need 147 Watt-hours.

So, in reality, the relative importance of natural gas and power in the home need to be essentially reversed, both for dollar and carbon considerations (Figure 4.5).

What's the Carbon Balance?

Another factor to consider is the amount of carbon dioxide emitted by each energy source. During combustion, oxygen combines with the fuel's hydrogen or carbon to release energy. However, only carbon

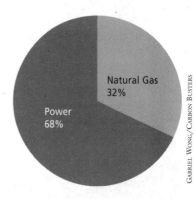

4.5: *Carbon Emissions of a typical US house (2,000-2,500 sq. ft.).*[24]

combines with oxygen to create carbon dioxide (CO_2); hydrogen and oxygen only create water. Therefore, the higher the ratio of carbon in your fuel, the more carbon dioxide is created. For example, coal is almost pure carbon, creating the highest CO_2 emissions per kWh. Natural gas is essentially methane, with the chemical formula CH_4. Therefore, for every atom of carbon there are four atoms of hydrogen providing energy during combustion, leading to lower CO_2 emissions than coal.

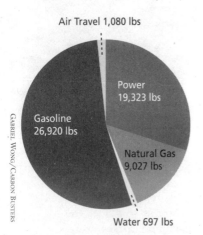

Air Travel 1,080 lbs

Power 19,323 lbs

Gasoline 26,920 lbs

Natural Gas 9,027 lbs

Water 697 lbs

GABRIEL WONG/CARBON BUSTERS

4.6: *Annual carbon dioxide pie of a typical US family living in a 2,000 to 2,500 sq. ft. home, with one sedan and one SUV (about 57,000 lbs).*[25]

This means that when using natural gas at a generating plant, only about a third as much CO_2 is produced per kWh compared to a plant that uses coal for power generation (check Appendix I on page 146 to find out about your area).

This explains why using coal-generated power to heat your kettle of water can create five times as much CO_2 as heating it with natural gas directly, as illustrated in our earlier example (Fig. 4.4).

How to Benefit from Fuel Switching

Because of the lower cost of using primary energy directly for heating purposes rather than secondary energy like electricity, using natural gas in this case saves you money as well as dramatically lowering your CO_2 emissions (by as much as 80 percent).

So, if you live in a state or province that uses a lot of coal during power production, e.g. Wyoming or Alberta, definitely consider switching to natural gas. Even in most other states fuel switching makes sense.

The best targets for this move are: home heating, cooking and clothes dryers.

Two down sides of natural gas are that the cost of natural gas is likely to rise more than the cost of electricity in the next few decades, due to limited supply (you have been warned!), and we are as yet unsure about the level of fugitive emissions of methane (a greenhouse gas) from natural gas pipelines.

Nevertheless, based on the current US national average, natural gas costs half as much and releases five times less CO_2 per kWh than coal-generated electricity when used in heating applications.

Your Home's Carbon Pie

Take a look at the carbon emissions in Figure 4.6.

Figure 4.6 displays the amount of CO_2 emitted by a typical 4-person family living in a detached house, with one sedan and one SUV, and traveling 270 miles via airplane per person per year. The total emissions amount to about 57,047 pounds per year!

Which is the Greatest Source of Carbon?

Quick Fact

Cars are your family's number one carbon emitters.

It may surprise you that your car (or cars) actually emits more CO_2 than any other residential source, around 27,000 pounds per year. This gives you some indication that an effective carbon busting strategy must involve a focus on transportation.[26]

The next highest source of carbon pollution caused by a typical family is electric power with 19,000 pounds.

Natural gas (used for space heating and domestic hot water) accounts for an additional 9,000 pounds.

Domestic flights create another 270 pounds based on 877 miles of average air travel per person.

Water consumption accounts for about 700 pounds of CO_2 emissions.

How Much Does it Cost?

Now let's examine the monetary cost of a typical family's energy bill.

Not surprisingly, gasoline costs for your vehicle(s) top the chart again, with $3,839 or 62 percent of the total energy bill (Fig. 4.7).

Natural gas is next, with $1,162, or 19 percent of your energy costs. In this scenario, power costs you $1,019 (16 percent), and water $219, or 3 percent.

Altogether it adds up to $6,239 each year. And this does not include fixed costs for your utilities (i.e., standard charges that are usually applied regardless

From Carbon to Carbon Dioxide

Beware when comparing figures on greenhouse gas emissions: are you dealing with "carbon" or "carbon dioxide"? And what's the difference?

When a fuel containing carbon gets combusted, each carbon atom combines with two oxygen atoms to form one molecule of carbon dioxide, with the chemical symbol of "CO_2".

You may recall that carbon has an atomic weight of 12, oxygen an atomic weight of 16.

So if you start out with one atom of carbon with an atomic weight of 12, you end up with a molecule containing one atom of carbon (12) and two atoms of oxygen (16 each), for a total atomic weight of 12 + 16 + 16 = 44.

The ratio of the final carbon dioxide to the initial carbon is 44 divided by 12, which is about 3.67. This means that for every pound of carbon burned, about 3.67 pounds of carbon dioxide result.

1 pound carbon = 3.67 pounds carbon dioxide

There! And you thought taking high school chemistry wasn't useful!

of actual consumption) or travel costs. Note: For a more detailed break-down of typical costs, see Appendix F. For US national average energy prices used as a basis for all cost calculations, see Appendix G.

We often think of buying a home as the biggest purchase of our life. How does the family energy cost compare to the purchase of a house over its lifetime (Fig. 4.8)?

In 2005, the average US cost of a single-family house with 2,200 sq. ft. and four bedrooms in "typical, middle-management neighborhoods" in one study was $401,767 (excluding interest costs).

If we assume a life-span of this house of 80 years, the energy cost (including gasoline) for the family living in the house is just under $500,000, or about 20 percent more than the cost of the house itself.

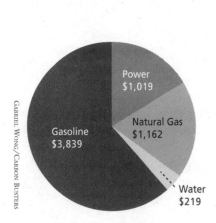

4.7: The annual energy bill of a typical US family is $6,239 (excluding air travel costs and fixed utility charges).[27]

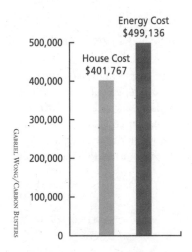

4.8: Purchase cost of a 2,200 sq. ft. single-family home vs. lifetime energy cost (assuming life of 80 years, including gasoline costs, and excluding interest payments and fixed utility costs, in constant 2005 dollars).[28]

5

Home Improvement:
Green Home Design

THE MAIN FOCUS OF THIS BOOK is how to reduce the consumption of your existing home, rather than building a new one (this may be the topic for a future work!).

However, knowing how innovative approaches in new buildings are dramatically reducing energy costs may inform and inspire your approaches to your own home. What are some ingenious ways in which people around the world are addressing home energy efficiency?

The Law of Diminishing Returns

Our energy efficiency efforts used to be ruled by the "law of diminishing returns" (Fig. 5.1). Put simply, this rule states that as spending increases, for each new dollar spent there is less and less of a return.

Let's take a simple example from home insulation: Let's say it costs $4,000 to insulate your house enough to cut your heating bill in half from $1,000 to $500. If you spend another $4,000 on insulation you may halve your fuel bill again, but one half of $500 is merely $250 in savings. And on it goes. How can the limitations of this "law" be overcome?

You see things; and you say, "Why?" But I dream things that never were; and I say, "Why not?"
— George Bernard Shaw, *Back to Methuselah*

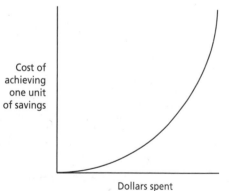

5.1: *The law of diminishing returns: As expenditure increases, the return per invested monetary unit declines.*

Tunneling Through the Cost Barrier

Here, think-tank Rocky Mountain Institute in Colorado comes to the rescue: the answer is a whole-system, highly integrated design approach that they call "tunneling through the cost barrier" (Fig. 5.2).

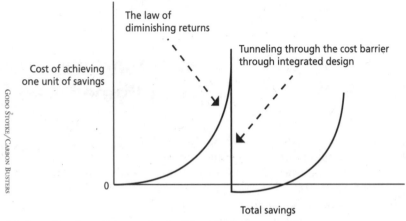

5.2: *Tunneling through the cost barrier (after RMI).*[29]

This approach looks not just at how savings can be achieved in one particular area (e.g.: heating or ventilation), but how one feature can benefit the total design, for example through savings in energy costs *as well as* equipment costs, or structural integrity, or thermal mass, etc. The approach optimizes the whole system and examines all resulting benefits.

Let's continue with the example of insulation: as we add more and more insulation, the return per dollar of insulation declines. What if we insulate the house so well that it no longer requires a furnace for heating or air conditioners for cooling?

That's exactly what the city of Lindås in Sweden did when they built 20 terrace houses south of Göteborg: they added excellent windows, better insulation, and an air-to-air heat exchanger. [30] These features added about $7,000 to the cost. However, they were able to dispense with the heating system altogether, at a savings of $4,000.[31] This leaves a net cost of $3,000 for the extra insulation, with a simple payback of under four years, and free heating for the remainder of the life of the house. (You can access a free English-language PDF of the design at www2.ebd.lth.se/avd%20ebd/main/Gothenburg/Folder_Lindas_EN.pdf.)

Whole-System Design

Let's look at a few other examples of successful whole-system design. Here is one closer to home:

Rocky Mountain Institute

The Rocky Mountain Institute (RMI) headquarters building was completed in 1984. It serves as the residence of RMI co-founder Amory Lovins, and as office space for some of RMI's 40 staff. Even though the building is over 20 years old, it is 99 percent solar heated (Fig. 5.3). Its power consumption is 90 percent lower than conventional buildings with comparable usage, and all of the power is supplied by photovoltaic panels through a grid-connected, net-metered system. Water consumption has been reduced by over 50 percent.

The 4,000 square foot structure, despite being located in subarctic temperatures at an elevation of 7,100 feet in the Colorado Rockies, has no furnace. Heating needs are greatly reduced by massive insulation; R-80 in the ceiling, and R-40 in the walls, about twice the standard for well-insulated modern buildings. The building contains an effective thermal mass of 250 metric tonnes of material that helps even out temperature. Most of the thermal mass is provided by the walls, which consist of two layers of local sandstone rocks infilled with mortar, sandwiching four inches of polyurethane foam for insulation.

Windows ranged initially in insulating value from R-4 (U-value of 0.25) to R-8 (U-value of 0.125), but

> **Quick Fact**
>
> In 1984 the Rocky Mountain Institute (RMI) in Colorado built a house that cut water consumption by 50 percent, power consumption by 90 percent, and heating by 99 percent, with a payback of ten months.
>
> "Today we could do better," says RMI co-founder Amory Lovins.

5.3: *The Rocky Mountain Institute headquarters building in Old Snowmass, Colorado.*

ROCKY MOUNTAIN INSTITUTE

were upgraded in the early nineties to up to R-12 (U-value of 0.083; for comparison, double-glazed windows are about R-2, or a U-value of 0.50). These superwindows consist of two glass panes and two Heat Mirror films, filled with krypton gas. Today, even higher insulation values in windows are possible (see Chapter 8, "Windows", page 109).

Due to the greatly lowered heat loss of the building, all the heat required can be supplied by solar gain through the windows, and heat supplied by people, pets and electrical equipment.

The central greenhouse with its large, super-efficient windows, takes on the role of the "furnace" for the building.

In addition, the building contains numerous advanced and innovative ways to reduce power and water consumption, some of which we will return to in later sections of the *Carbon Buster's Handbook*.

To take a virtual tour of the Rocky Mountain Institute go to www.rmi.org/sitepages/pid623.php. The most recent version of the visitors' guide can be downloaded for free at www.rmi.org/sitepages/pid379.php

What was the payback of the energy saving features?

The total construction cost was slightly over $500,000 in 1984 dollars, or around $130 per square foot (building costs in the Aspen area are about twice the national average). All the energy and water saving features increased the net cost by $6,000, or $1.50 per square foot (just over one percent). Compared with local building practice and usage, the building saves $7,100 per year. This means that the extra cost of the energy features paid for themselves in about ten months.

And, in typically cheerful manner, Amory Lovins adds that with recent advances in technology "one could do better today," and that "by 2054 the building's energy savings will have repaid its entire construction cost." Not bad.[32]

Alberta Sustainable Home

The Alberta Sustainable Home (ASH) is a Calgary house designed by architect Jorg Ostrowski and friends, and built in 1993 (Fig. 5.4). It reduces energy consumption by an average of 94 percent, going

GODO STOYKE/CARBON BUSTERS

5.4: *The Alberta Sustainable Home in Calgary, Alberta (solar oven in foreground).*

from a purchased energy consumption of 68,000 Btu per square foot a year to just 4,000 Btu/sq. ft. per year.

The 1,800-square-foot building has walls with an insulation value of R-50, and a roof with a value of R-70, provided by blown cellulose. The walls are 14 inches thick, and are made of "eco studs", recycled framing materials that consist of a 2 by 4 stud connected to another 2 by 2 held apart by 2 by 4 spacers (Fig. 5.5). This design not only creates a wider wall allowing the insertion of more insulating material, but also reduces the thermal bridging — a result of all the 2 by 6 studs in a conventionally framed house — by 95 percent.

Windows range from R-5 (U-value 0.2) to R-8 (U-value 0.13), as well as a superwindow prototype with an R-value of 17 (center of glass; U-value 0.06).

The house has no connection to natural gas, or any other fossil fuel for heating. Instead, hot-water solar vacuum tubes supply heat to the house. A masonry stove (mass stove) can be used for backup heat, as well as for cooking, if desired. Most of the cooking, though, is done in a solar oven on the veranda. (Jorg uses two types, one from Saskatchewan and one from Switzerland. The Swiss model is so effective that it may actually get *too* hot.)

5.5: *Raising a wall made of eco-studs.*

Another unique feature of the house is that it is not connected to the city mains, deriving all its water from precipitation. The house has one advanced composting toilet (the Phoenix) and one ultra-low flush toilet that was actually developed for RVs (the Sealand). The Sealand also feeds into the composting toilet. Grey water (water from shower, bath and kitchen) is recycled and treated inside the house.[33]

International Netherlands Group (ING) Bank, Amsterdam

The headquarters of the Inter-national Netherlands Group (ING) Bank in Amsterdam is another showcase of successful energy-efficient design (Fig. 5.6).

5.6: *The International Netherlands Group (ING) Bank in Amsterdam achieved a 92 percent reduction in energy consumption compared to its previous building.*

Quick Fact

The new ING Bank headquarters in Amsterdam cut its energy consumption by 92 percent. However, increased staff productivity and reduced absenteeism, due to the green building design, saved the bank 30 times more money than the reduced energy costs.

The building was so innovative that some even credit it with its owner's turnaround from Holland's fourth- to its second-largest bank.

The building's chief architect is Anton Albers, who was inspired by Rudolph Steiner, the inventor of Waldorf schools and bio-dynamic farming (a kind of organic, low-input agriculture). There are no right angles in the structure, art was integrated into the building during the design phase, and there are lots of plants and water-features present.

The building houses 2,400 employees and occupies over half a million square feet. The first step in the three-year design phase for the building, completed in 1987, was determining where its employees lived — then the building was sited to minimize total employee travel time.

The next step was a reduction in the energy required to heat and cool the building. As well as obtaining heat from passive solar design, the building incorporates air-to-air heat exchangers to pre-heat incoming fresh air during the cold season. In addition, hot air from computer rooms and elevator motors, and waste heat from a cogeneration plant that produces electricity warm the building, thereby greatly increasing the total efficiency of the design.

The building is supplied with ample windows, providing daylighting for the work-spaces. No office desk is more than 23 feet from a window, obviating the need for most artificial light.

In addition to the numerous windows, clerestories and atria, bright, reflective walls direct light throughout the building, even all the way to the basement. Most of the remaining light needs are met by task-lighting that provides light specific to each individual's needs and has low power requirements.

While the old building used 422 kBtu/sq. ft., the new one uses only 25 kBtu/sq. ft., a reduction of 92 percent (1,000 Btu equal one kBtu). Even a neighboring bank building, built for the same cost and at the same time, consumes five times more energy than the ING headquarters.

The energy efficiency measures altogether cost an additional 1.5 million guilders ($876,000). However, due to the annual energy savings of 5 million guilders ($2.9 million), the payback for all measures was under four months.

The greatest surprise, however, came with the evaluation of staff productivity in the new headquarters. Due to the improved ambiance and work environment, ranging from natural daylighting, reduced mechanical noises, and improved air quality, absenteeism and work lost to sick-days was dramatically reduced. In fact, the increased work hours were worth 30 times as much as the energy savings themselves.[34]

Green Transportation

TRANSPORTATION, SPECIFICALLY OUR CARS, represents the biggest slice of our annual energy bill, and of our carbon pie: $3,839 and 27,000 lbs; (53 percent), respectively (see Figs. 4.6, and 4.7 in Chapter 4: "Your Family's Carbon Pie"). So this is a good place to start cutting carbon. This chapter examines North America's most energy-efficient cars, and some car alternatives.

Efficiency Comparisons

How efficient is our transportation system? While riding a car, only 1 percent of the energy contained in gasoline is actually used to move the passengers.[35]

How does that compare to other modes of transportation? Fig. 6.1 compares the energy efficiency of planes, cars, buses, trains, walking and biking. Planes and cars are the least efficient modes of transportation, with planes nudging out the car in efficiency over distances of 600 miles, and vice versa for distances under 600 miles. Buses tend to be twice as efficient as cars, trains three times as efficient. A bike is twice as efficient as walking, and 50 times more efficient than a car.

Bicycling

Cars are generally assumed to be faster than bicycles, and for long distances this holds true. For shorter distances, such as daily commuting, the situation is often reversed. An important factor is the amount of traffic: for example, in Edmonton's annual commuter race (a bike, a bus, a car), the bike wins every time. The other factor is the cost. The annual cost of maintaining a car

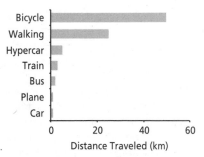

6.1: *Distance traveled per 1,100 calories of energy (50 km equals 32 miles).*

Quick Fact

Bicycles are the most efficient vehicles on the planet, 50 times more efficient than cars, and twice as efficient as walking.

is around $7,000. If you earn $14 per hour after tax, it takes you three months each year just to pay for your car, compared to $220, or two days, for your bike. If you add this work time to your driving, the average speed of your car just dropped quite a bit. Or, as the saying goes, "Drive to Work, Work to Drive." Therefore, if you have flexibility in how much time you work per day, working fewer hours each day and biking instead may be more efficient for you.

For some interesting tips on "How to Not Get Hit by a Car", check out bicyclesafe.com.

Public Transit

Depending on where you live, the cost of an annual public transit pass may be in the range of $696 (Los Angeles County) to $912 (New York City), compared to $7,000 for the annual cost of maintaining a car. If you use some of the saved money to pay for a holiday trip rental car, and the occasional cab, you still come out ahead by a wide margin.

By using a bus instead of a car, you will cut your transportation carbon emissions in half, by using a train or subway, it will drop to one third.

You can also persuade your employer to offer discounted or free transit passes to all staff. Many cities offer special group discount rates to companies for their staff. The companies can either pass on the savings straight or add their own incentives.

"If you live in a city, you don't need to own a car."

— William Clay Ford Jr., CEO, Ford Motor Company[36]

Cars

Carsharing

Carsharing is based on the premise that many of us can fulfill most of our transportation needs with walking, biking and public transit. For those times when we can't, we can rent a car, or join a carsharing service — over 72,410 Americans and Canadians already do so.

There are a number of fixed costs for owning a car, such as loan or lease payments, insurance, registration and, in some areas, paying for a parking stall. If you don't drive very much, but still want the convenience of an automobile for specific jobs, carsharing may be the answer. According to the folks at carsharing.net, it is an economical alternative for those driving 7,500 miles a year or less. The next question is whether there is a depot near your home. Use carsharing.net to find out where your nearest carsharing depot is located, or in Canada check out carsharing.ca and location listings at autoshare.com.

Carpooling Incentives

An increasing number of states are offering express lanes for those who carpool, or even for approved low-emissions vehicles like hybrids (these require a sticker issued by the state). To find out who offers or needs a ride to your destination in Canada and the US try rideshare.us or erideshare.com, in Canada you may also want to go to carpooltool.com. There are also commercial services such as carpoolworld.com and others which you can find with an Internet search.

Hypercars

The Hypercar is a concept that was introduced by the Rocky Mountain Institute's Amory Lovins in 1991. It represents the radical redesign of the car for greater efficiency from the tires on up.

Change is inevitable. Except from a vending machine.

— Robert C. Gallagher

Eighty-five to 87 percent of the energy contained in a conventional car's fuel is lost before it reaches the tires, and only 1 percent is ultimately used to move the passengers.

What causes the high consumption of conventional cars? The most important aspect is weight – two-thirds to three-quarters of the energy consumption is due to the car's weight, which in turn requires a larger, heavier motor and a heavier superstructure.

One of the most important innovations of true hypercars therefore is reduced weight due to advanced carbon-fiber composites, such as carbon-fiber reinforced polymer (C-FRP). Such materials are already being used successfully, for example in Formula One race cars, where drivers are rarely killed despite speeds of 200 mph. Pound for pound, these carbon fibers can deliver six to twelve times the strength of steel (Fig. 6.2). While carbon fibers are more expensive than steel, part of the higher cost is offset by the need for less material in the final car. Carbon fiber can also reduce the total number of auto parts required. A typical car can consist of about 1,000 parts. The design and production of the dies for the parts may have a cost of $1 million per part, for a total production-line development cost of around $1 billion. In hypercars, car parts may be reduced to 300 or fewer, greatly lowering production-line development costs.

Another trick to making hypercars more efficient is to reduce air drag. This is most easily accomplished by making the underside of the vehicle as smooth as the top, since, as Amory puts it, "the air doesn't know which side it's on."[37]

One example of the hypercar is the Revolution SUV virtual design by the non-profit Rocky Mountain Institute's for-profit spin-off Hypercar, Inc. (now Fiber Forge), a safe 99-mpg mid-size SUV that provides seating for five (Figure 6.3).[38]

Hypercars are not available for sale today, but there are a number of existing prototypes that come close. For example, Opel, the European subsidiary of GM,

	Standard	Hypercar
Fuel efficiency	21 mpg	78-235 mpg
Construction material	Steel	Carbon fibre composite
Crash absorption (relative to steel=1)	1	6-12
Aerodynamic drag coefficient (Cd - typical value)	0.3	0.18
Regenerative braking	No	Yes
Cheaper	No	Yes
Safer	No	Yes
Faster	No	Yes
Range 550-850 miles	No	Yes
4-5 passengers	Yes	Yes
No standby losses	No	Yes
Faster manufacturing	No	Yes
More durable	No	Yes

6.2: *Key characteristics of regular cars and hypercars.*[39]

6.3: *The "Revolution" is the Rocky Mountain Institute's stab at producing a concept study for a roomy, peppy, safe, and inexpensive SUV that achieves 99 mpg.*

Quick Fact

GM Opel's working Eco-Speedster prototype has a top speed of 155 mph, and a fuel efficiency of 94 mpg.

has created the "Eco-Speedster", a sports- car with a top speed of over 155 mph. Despite its 117 hp motor, the car gets 94 mpg. The 1,450 pound car has a four-cylinder 1.3-Liter-CDTI common rail turbo diesel engine and a sleek carbon fiber body with an aerodynamic drag coefficient of only 0.2. It seats two (Figure 6.4).[40]

Another "almost" hypercar driving on Europe's roads today (but not available for sale anywhere) is Volkswagen's "1-Liter-Auto". The "1-Liter" tag refers to its rated one-liter consumption per hundred kilometers, though on its maiden voyage it actually got an astonishing 264 mpg.

This street-legal car, also a two-seater, has a carbon-fiber body that weighs only 640 pounds. It is propelled by a 1-cylinder, 300cc diesel engine that delivers 8.5 hp, yet reaches a respectable top speed of 75 mph. The diesel car also has regenerative braking, feeding deceleration energy into Nickel-metal-hydride batteries (NiMH), a feature it shares with today's hybrid cars (see below). The tank is designed for robotic refueling and holds a mere 1.7 gallons, yet the car still has a range of 450 miles. Despite its low weight, the vehicle has ABS brakes, and a front-seat airbag. Deformation elements in the front, crashtubes and the magnesium spaceframe construction ensure that the 1-Litre-Auto car has the equivalent crash-protection of a GT-class race car.

The daytime running, signal and taillights of the 1-Litre car are energy-efficient LEDs, with a 30 W bi-xenon bulb providing bright nighttime viewing equivalent to a 60 W conventional bulb. The car has no side mirrors, to reduce air drag. Instead, it relies on CCD cameras and in-dash monitors to provide information on the proximity of other cars, as well as a camera mounted on the brakelights for backing up. The car's shape is more reminiscent of a glider airplane than a car (Fig. 6.5).

Unfortunately, VW has no plans to build this car, saying that it is "too expensive" to build (around $20,000 per car).[41]

Except for the number of passengers, these cars come very close to RMI's "hypercar" concept, yet they already exist and have been running legally on streets for a number of years now. It is frustrating that none of the large car manufacturers has produced hypercars in production quantities yet. I would

ADAM OPEL GMBH

6.4: *The Opel* Eco-Speedster, *with a top speed of 155 mph — more than adequate to get to work on time — is proof that exceptional speed and exceptional energy efficiency (94 mpg) are not incompatible.*

Quick Fact

VW's street-legal 1-Litre-Auto carbon-fiber car prototype gets 264 mpg, yet has the crash-protection of a GT-class race car.

VOLKSWAGEN AG.

6.5: *Volkswagen's* One-Litre-Car *achieved 264 mpg on its Autobahn maiden voyage.*

not be surprised if the well-known car companies pay the financial penalty sometime in the near future and get eclipsed by an as-yet totally unknown maverick company that will embrace a radically redesigned approach to hypercar manufacturing. Until then, let's take a look at what is actually available on the market today.

Top Carbon-Busting Cars

Here are the three most important things you can do to decrease the energy consumption of your car:

1. Pick the right car
2. Choose the right car, and
3. Yes, you guessed it: buy the right car.

There is nothing wrong, of course, with checking your tire pressure, and avoiding wicked starts and brutal braking to reduce fuel consumption. However, that is going to be peanuts compared to the effect your choice of vehicle has.

Since the average car ownership period is 11 years, this is a decision you may live with for a long time. Or, you may simply want to trade in your car and start saving now.

How do fuel costs and car purchase cost compare?

Let's start with a car whose energy consumption is relatively high compared to its purchase price. The 2006 Chevrolet Equinox has a manufacturer's suggested retail price (MSRP) of $ 21,955. We'll assume that the car is driven for 200,000 miles (a realistic figure for the Honda and Toyota hybrids we will be comparing this with, though perhaps not for the base case. In that case, you would have to add the additional car purchase costs to the base case compared to the green cars). Driving this distance in the Equinox, you will pay $ 28,426 for gasoline at constant national average US gasoline costs at the time of writing. In other words, the purchase cost, high as it may seem, is actually eclipsed by the total fuel costs (Fig. 6.6).

Considering how much fuel costs add to the life-cycle cost of your car, it is surprising how pitiful the fuel efficiency knowledge of many car salespeople is.

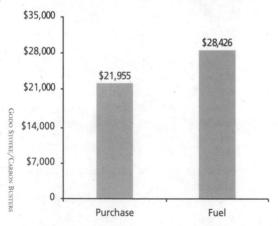

6.6: *Purchase and lifetime fuel costs of a 2006 Chevrolet Equinox.*

If we compare the purchase and fuel life-cycle cost with an equally expensive but more efficient car, such as the Toyota Prius, the difference that fuel efficiency can make in the sum of these costs is quite clear: The cost for the Equinox is $ 50,381, while the costs for the Prius, with about the same initial purchase price, add up to only $32,352, a fuel savings of about $17,800 (Figure 6.7).[42]

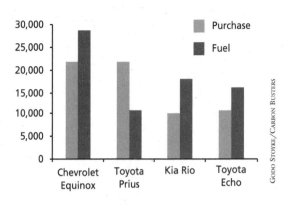

6.7: *Total cost of car purchase and fuel.*

What are the most efficient cars? Take a look at the list on pp. 38-39.

The most fuel-efficient *used* cars you can buy are the Geo Metro (45 mpg) — also sold as the Pontiac Firefly, Chevy Sprint, and Suzuki Swift, and the Honda Civic HB VX, which gets 51 mpg, and was available new from 1992 to 1995. From 2006 on, we can add the Toyota Echo to the used category. Until now, the Echo was our most fuel-efficient gas-powered car (as opposed to gas-electric or diesel), but has been replaced in the US by the slightly less efficient Toyota Yaris.

Access: To download the latest copy of the US Government Fuel Economy Guide, go to fueleconomy.gov. For the Canadian edition, go to vehicles.gc.ca. The most efficient vehicles in each category are listed at edmunds.com, and in Canada at vehicles.gc.ca.[43]

Gas-Electric Hybrids

You will notice that three of the top four cars are hybrid vehicles. How do hybrids achieve such amazing efficiencies? These "gas-electric" hybrids have electric motors in addition to traditional gasoline-powered internal combustion engines. The presence of the electric motor has several advantages:

1. The electric motor assists in acceleration, allowing the gas motor to be down-sized, which reduces fuel consumption during operation. (Conventional car motors have to be strong enough to allow for rapid acceleration, e.g.: when passing other cars, and to climb hills. The large engine leads to inefficiencies during normal city and highway driving, when only a fraction of the engine's power is needed.)

2. The battery system propelling the electric motor can accept kinetic energy during braking ("regenerative braking"), greatly reducing the stop-and-go losses found especially in city driving.

3. The electric engine can take over completely at many urban driving speeds, allowing the gas engine to be shut off completely (though not in all hybrids).

4. The gas engine is also shut off as soon as the foot is taken off the accelerator, and when the engine comes to a stop, eliminating idling losses.

5. Unlike those in electric cars, the battery system is relatively small, lightweight (e.g.: 99 pounds in the Prius), and inexpensive, since they are not the sole propulsive system.

6. Unlike electric cars, gas-electric hybrids never have to be plugged in.

Greenest Cars

	Mixed (55/45) MPG[1]	City MPG	Highway MPG	City L/100 km	Highway L/100 km	Mixed (55/45) L/100 km
Honda Insight	62.7	60	66	3.9	3.6	3.8
Toyota Prius	56.0	60	51	3.9	4.6	4.2
Mercedes Smart[2,3]	56.0	51	62	4.6	3.8	4.2
Honda Civic Hybrid	49.9	49	51	4.8	4.6	4.7
Volkswagen Golf TDI[3,4]	43.9	38	51	6.2	4.6	5.5
Volkswagen New Beetle TDI	40.2	37	44	6.4	5.3	5.9
Toyota Camry Hybrid	39.1	40	38	5.9	6.2	6.0
Volkswagen Jetta TDI	38.3	36	41	6.5	5.7	6.2
Toyota Echo	38.2	35	42	6.7	5.6	6.2
Toyota Yaris[3]	38.0	34	43	6.9	5.5	6.3
Toyota Corolla	36.1	32	41	7.4	5.7	6.6
Honda Civic	34.5	30	40	7.8	5.9	7.0
Scion xA	34.3	32	37	7.4	6.4	6.9
Hyundai Accent	33.4	32	35	7.4	6.7	7.1
Kia Rio	33.4	32	35	7.4	6.7	7.1
Volkswagen Passat TDI[3,5]	33.4	27	41	8.7	5.7	7.4
Large Cars						
Hyundai Sonata	28.5	24	34	9.8	6.9	8.5
Small Station Wagons						
Jetta TDI Wagon	38.3	36	41	6.5	5.7	6.2
Honda Fit	35.3	33	38	7.1	6.2	6.7
Midsize Station Wagons						
Ford Focus Station Wagon	29.6	26	34	9.0	6.9	8.1

The combined efficiency of the gas-electric system greatly increases the range of the vehicle over gasoline-only or electricity-only cars, which means you don't have to refuel as often.

Honda Insight

The Honda Insight is the single most energy-efficient car in North America, with an average fuel efficiency of 63 mpg (Figure 6.8).

The Honda Insight achieves its high fuel efficiency through its low weight, aerodynamic body and hybrid gas-electric engine. Its low aerodynamic drag, 0.25, the lowest drag of any mass-produced car in the world, is achieved through its

	Mixed (55/45) MPG[1]	City MPG	Highway MPG	City L/100 km	Highway L/100 km	Mixed (55/45) L/100 km
Standard Pickup Trucks						
Ford Ranger Pickup 2WD	26.3	24	29	9.8	8.1	9.0
Mazda B2300 2WD	26.3	24	29	9.8	8.1	9.0
Cargo Vans						
Chevrolet G1500/2500 Chevy Express 2WD	17.3	15	20	15.7	11.8	13.9
GMC G1500/2500 Savana 2WD	17.3	15	20	15.7	11.8	13.9
Passenger Vans						
Chevrolet G1500/2500 Chevy 2WD	16.8	15	19	15.7	12.4	14.2
GMC G1500/2500 Savana 2WD	16.8	15	19	15.7	12.4	14.2
Minivans						
Honday Odyssey 2WD	23.6	20	28	11.8	8.4	10.2
SUV						
Ford Escape Hybrid	33.8	36	31	6.5	7.6	7.0
Best Used Cars						
Geo Metro	45.1	43	48	5.5	4.9	5.2
Honda Civic HB VX	51.2	48	55	4.9	4.3	4.6

[1]Based on most recent available data, including 2005, 2006 and 2007 models. [2]Available in US through ZAP cars only (expensive!). Expected availability in US through Mercedes: 2007; [3]Fuel efficiency based on Canadian Government test standards, which deviate slightly from US EPA standards; [4]Available only in Canada; [5] Only available in Canada until 2005.

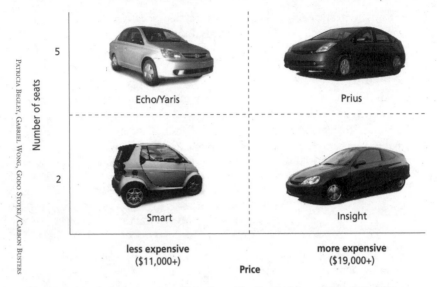

6.8: *North America's most energy efficient cars. The Toyota Echo and Honda Insight were discontinued in 2006, but may be available new from dealer stock, or purchased used. Alternatives to the cars shown include the Honda Civic Hybrid, VW Jetta TDI, New Beetle TDI, Golf TDI and Toyota Camry Hybrid.*

teardrop shape and smooth underbody covers, which ensure clean airflow under the car, realizing yet another feature of RMI's "hypercar" concept outlined above.

The Insight weighs 1,823 pounds, and has a 67-hp 1-liter, 3-cylinder gasoline engine, and a 14-hp electric motor, for a combined yield of 79 hp. Since most commutes involve only one driver, Honda decided to make the Insight a two-seater. To further reduce weight, the body is made from aluminum, thereby also improving its crash-worthiness. The aluminum reduces body weight by 47 percent, yet increases rigidity for better handling.

The Insight's engine automatically shuts off instead of idling.

The base price of the manual Honda Insight is $19,330. The automatic version has a base price of $21,530 and comes with air conditioning. Ironically, the US federal hybrid car incentive of $1,450 appears to be available only for the less-efficient automatic transmission version.[46]

Driving an Insight as your secondary car, you can reduce your car's lifetime CO_2 emissions by 88,000 pounds, and save over $12,500 in fuel, compared to a 24 mpg car.

Quick Fact

The Honda Insight, with 66 mpg (highway), is North America's most fuel-efficient car.

Common Hybrid Myths

Do I have to plug in my hybrid car?

No. Hybrid car batteries are recharged entirely by braking energy in combination with the internal combustion gasoline engine, and never have to be plugged in. (However plug-in hybrid advocates, e.g. calcars.org, have converted regular hybrids to plug-in hybrids — PHEVs — to significantly increase mileage.)

Do I have to replace the batteries every few years?

No. The Honda and Toyota hybrid batteries are designed to last for the life of the vehicle.

Will hybrids reach the EPA-rated mileage?

No. EPA MPG ratings are under ideal conditions and will not be reached by the hybrids in real life (one of the reasons the EPA is revising their mpg rating standards in 2006). However, the same applies to all other cars rated under the system. Therefore, EPA mileage ratings are still a good way to determine the relative fuel efficiency performance of cars, including hybrids.[44]

Recommendation: Replace your secondary car with a Honda Insight (since the Insight is a two-seater, replacing your primary car with an Insight will not be practical for most families. However, if there are only one or two people in your household, you can achieve even greater savings by making the Insight your primary car (see below).

5-year savings: $3,821, 27,000 lbs. CO2, 1,370 gallons gasoline (all 5-year car savings are based on driving your primary vehicle 15,000 miles in the first, and 12,000 in each of the following years, and your secondary vehicle 12,000 miles per year. The assumed fuel efficiency for your current vehicles is 17 mpg for your primary vehicle, and 24 mpg for your secondary vehicle).

Life-time savings: $12,546, 88,000 lbs. CO2, over 200,000 miles.

Incremental cost: $3,000 ($1,550 for manual version).[45] New cost: $19,330. An estimated $1,450 federal tax credit is available to the first 60,000 purchasers of Honda low-emission vehicles (automatic drive only). The automatic drive (CVT with air conditioning) costs $21,530, or $1,000 more than the manual version with air conditioning, $20,080 after the tax credit. You will lose about $330 in fuel savings every five years if you choose the CVT version.

Simple payback: 3.9 years (2.2 years for the automatic shift version, due to government tax credit), IRR: 24.7 %, CROI: 29.3 lbs./$. (See page 11 or the glossary on page 150 for an explanation of the acronyms and other technical terms.)

Note: If your Insight replaces your primary car (17 mpg), your lifetime savings are $22,576, and 158,000 lbs. of CO2, i.e., the Insight will save you more in fuel costs over its life than its purchase price, even without counting the hybrid rebate. See p. 140 for details. Due to the relatively low volume of Insights sold per year, the Insight will be discontinued in September of 2006.

Toyota Prius

The Toyota Prius was first produced in 1997, but was only sold in Japan — it came to the US for the first time in 2000, as the first hybrid with more than two seats (the Honda Insight was the first hybrid in the US).

The 5-seater Prius has been a runaway success, rising in global sales from 300 in 1997, to just under 200,000 units in 2005 (Figure 6.9).

Even now, demand often exceeds supply, leading to wait-times of six months in California. The Prius has excellent resale value — it is one of the few cars that has repeatedly been reported to have sold for more than its list price in used condition.

The Prius is now in its third iteration, and the latest version added an increase of 20 percent in interior space, while at the same time decreasing consumption by a further 15 percent. Toyota has a very advanced algorithm that controls the interplay of the electric and gasoline engines, reducing fuel consumption, particularly in city driving where the Prius excels. Several other car manufacturers are licensing Toyota's software.

The Prius is the most energy efficient car with more than two seats in North America — it seats five comfortably and is the first energy-efficient hybrid that offers a serious alternative to SUVs in terms of space for all but very large families.

 Access: toyota.com/prius, en.wikipedia.org/wiki/Toyota_Prius.

Carbon Buster Recommendation: Replace primary family car (17 mpg) with a Toyota Prius.

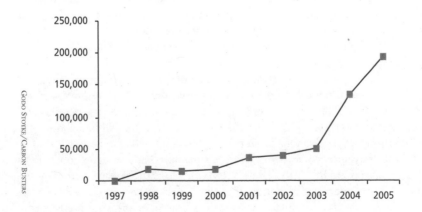

6.9: *Global Toyota Prius hybrid sales.*

According to industry estimates, the hybrid component of the Prius costs about $3,000 more compared to the average car. On the other hand, the US federal government offered tax deductions of $2,000 in 2005, and is offering up to $3,150 in 2006 to 2010 (though the number of rebates is limited for each manufacturer). Based on this incentive, you are actually ahead $150. Incidentally, a number of states and organizations offer further tax incentives. For example, the state of Colorado offers up to $3,285 in tax rebates for a hybrid purchase. Software developer Hyperion and web-site search tool Google both offer their employees an additional bonus ($5,000 and $1,500 respectively) for a hybrid purchase. However, "mild" hybrids that only marginally raise fuel efficiency do not qualify. The minimum efficiency is 45 mpg, i.e. the Honda Insight, Toyota Prius and Honda Civic Hybrid purchases are eligible. Based on the above amounts, the lucky Colorado Hyperion employees are eligible for total hybrid rebates and incentives of $10,860.[47]

> **Quick Tip**
>
> A Toyota Prius hybrid car will save you about $17,800 in fuel costs over 200,000 miles of driving compared to the Chevrolet Equinox (which has the same purchase price as the Prius).

5-year savings: $6,722, 47,000 lbs. CO2, 2,400 gallons gasoline.[48]

Life-time savings: $21,341, 149,600 lbs. CO2 over 200,000 miles.

Incremental cost: $3,000 for hybrid drive, minus $1,575 federal tax credit equals a net cost of $1,425. **New cost:** $21,725 minus $1,575 federal tax credit equals $20,150. Note: you may be eligible for significant further state tax credits or employer bonuses — see "Hybrid Tax Credits", p. 44.

Payback: 1.1 years. **IRR:** 94 %. **CROI:** 105 lbs./$.

If you don't need to replace your car yet, but would like to do so anyway to reap the environmental and economic benefits, how does this change the payback picture?

The difference basically consists of the penalty you incur by trading in your car, and the new registration costs. The penalty for trading in your car is dependent on the car age and depreciation rate (new cars losing more annual resale value than older cars). If we assume an average replacement penalty of $2,000, the payback e.g.: for the Prius, changes from 1.1 years to 2.5 years.

Honda Civic Hybrid

The Honda Civic Hybrid is a compact car with an average fuel efficiency of 50 mpg. For the first time, the 2006 Civic Hybrid is now able to run on the electric engine alone when less power is demanded, just like the Toyota Prius.

The Civic Hybrid is not quite as efficient as the Insight, Prius or Smart, but still 10 mpg better than the next closest competitors, the Volkswagen New Beetle TDI and Golf TDI.

Carbon Buster: Cumulative CO2 reduction 16.5%; 5 year savings $6,722.50

The Honda Civic Hybrid has a base price of $ 21,850.

Access: automobiles.honda.com, en.wikipedia.org/wiki/Honda_Civic_Hybrid.

Hybrid Tax Credits
FEDERAL INCENTIVES

Most hybrids are also eligible for federal tax credits in the US. There are currently no hybrid tax incentives from the Canadian government.

The US federal low-emission vehicle tax credit (up to $3,150 for hybrids, up to $4,000 for the natural gas Honda Civic GX, and up to $2,175 for the diesel Mercedes-Benz E320 BlueTec) is available from January 2006 to December 2010. The full credit is available for the first 60,000 qualifying cars of each manufacturer. Once the first 60,000 are sold, you can still get 100 percent of the credit for another three months, then 50 percent for half a year, and finally 25 percent for another six months before the credit for that brand expires.

By the time you read this book, Toyota is likely to have sold more than 60,000 of its popular Prius cars since the activation date of the incentive. Regardless of the hybrid you are purchasing, the faster you act, the higher your tax credit is likely to be. And remember, unlike a tax deduction, you will reduce your tax bill by the full amount of the tax credit. Check the following web sites for current credits:

Access: aceee.org/transportation/hybtaxcred.htm — Web site of the American Council for an Energy Efficient Economy, or download at aceee.org/transportation/taxcredits06.pdf

Hybrid Benefits

Sure, you can save up to $22,576 in gasoline costs by driving a hybrid. But there are numerous other benefits to being a hybrid driver:

- Free metered parking (e.g., Los Angeles, San Jose)
- Use of High Occupancy Vehicle (HOV) lanes for faster travel
- Up to a 50 percent reduction in the time and hassle spent refueling the vehicle
- Federal tax credits of up to $3,150 for buying a hybrid car
- State tax credits of up to $3,285 (e.g.: Colorado, Oregon) for purchase of a hybrid car
- Employer bonuses of up to $5,000 for purchase of a hybrid car (e.g.: Hyperion.com, Google.com)
- 10 percent discount on insurance rates for hybrid drivers (e.g.: Hybridtravelers.com)
- Reduction in emissions of the greenhouse gas carbon dioxide by up to 158,000 pounds over the life of the car.[49]

fueleconomy.gov/feg/tax_hybrid_new.shtml — Web site of the US federal government.

energytaxincentives.org — Web site of the Tax Incentives Assistance Project (TIAP), a US nonprofit coalition.

STATE INCENTIVES

For state and municipal incentives check out: eere.energy.gov/cleancities
(For example, Los Angeles and San Jose provide free parking at parking meters for hybrid vehicles.)

What About the Other Hybrids?

Hybrids are sprouting like mushrooms from the fertile soil of car showrooms, and that's a good thing. However, while they are significantly more efficient than their non-hybrid counterparts, none of the new models approach the fuel efficiency of the Honda Insight, Toyota Prius or Honda Civic Hybrid. Even the 2007 Toyota Camry Hybrid, with an average fuel efficiency of 39 mpg is still less efficient than the Volkswagen Golf TDI and the Volkswagen New Beetle TDI.

The Diesel Generation

Diesel cars have long been considered dirty, noisy and hard to start. However, great advances have been made in diesel engines over the last ten years. Today's best diesel engines can be equipped with particulate filters, starting is no longer a problem, even in very cold temperatures, and the engines consume far less fuel than gasoline engines. Another advantage of diesel engines is that they are the only type of engine that can run on biodiesel fuel (see "Biodiesel," page 49). An efficient diesel car running on biodiesel from a sustainable source beats even the best hybrid in terms of carbon emissions.

On the down side, diesel engines still give off more particulates, sulfur dioxide and nitrous oxides, and diesel produces 15 percent more CO_2 than gasoline, for the same volume of fuel.

Overall I'd say that hybrids trump diesels, and diesels beat regular gasoline vehicles. However, the world's most energy-efficient car, VW's "1-Litre-Auto" car, actually combines the benefits of diesel and hybrid engines, so my bet right now is on diesel plug-in hybrids as the most effective way of reducing greenhouse gas emissions in the near future, maybe followed by ethanol hybrids. Both the US (2006) and Canada (2007) have introduced, or are introducing, legislation to reduce the SO_2 content of diesel fuel, and existing diesels can be retrofitted with filters to reduce emissions at a cost of about $800.

In North America, the most energy efficient diesels widely available are Volkswagen's New Beetle, Golf and Jetta (38 to 44 mpg) at prices of $18,390, $19,580, and $21,605, respectively (the Jetta also has the distinction of having

Carbon Buster: 16.5%; $6,722.50

one of the lowest accident death rates of North American cars — see "Car Safety," p. 52).

An even more energy efficient diesel, rivaling the Prius in gas mileage, is the Smart ForTwo, manufactured by Mercedes (Figure 6.10).

The Smart car was designed by the maker of the Swatch wrist watch. And just as with the Swatch watches, you can actually get new face plates for the car to change its styling. The car was inspired by Bavarian ecologist Frederic Vester, who envisioned a car short enough to economically fit sideways on train cars, to reduce the impact of cars on long-distance travel. The design has definitely been successful in this category: at a length of just over 8 feet, the Smart is actually 4 feet shorter than the already-small BMW Mini Cooper.

It is the only car that legally parks sideways in parking stalls in Europe — two or three Smarts fit in a regular parking slot, which is one reason the Smart has been very successful in congested European cities, where parking space is at a premium. Despite its diminutive size, the two-seater Smart ForTwo offers plenty of space for two adults. Also, special attention was paid to safety considerations, including standard antilock brake systems and a tridion safety cell around the passengers, allowing the Smart to compete successfully with respect to safety with cars much larger than itself.

GODO STOYKE/CARBON BUSTERS

6.10: *Mercedes' Smart ForTwo Car is one of the cheapest vehicles in its efficiency class (city/highway 51/62 mpg), yet features many characteristics of a luxury car.*

Quick Fact

The Mercedes-Benz Smart, widely available in Canada and Mexico, and with limited availability in the US starting in 2006, is North America's most energy efficient diesel, and — arguably — its cutest mass-produced car.

In North America, the Smart is currently available from Mercedes only in Canada and Mexico, but it is expected that Mercedes will introduce it in the US in 2007. Alternative vehicle manufacturer Zap Zapworld.com has started shipping US-adapted Smart Cars in selected states, though at a premium. Check their web site for updates (according to the Zap website, ZAP is not affiliated with, or authorized by, smart gmbh, the manufacturer of SMART automobiles, or the smartUSA division of Mercedes-Benz LLC, the exclusive authorized US importer and distributor of those vehicles. ZAP purchases its vehicles from non-affiliated direct importer SmartAuto LLC.)

The Smart ForTwo currently sells for US$ 14,275 in Canada. If you think this a lot of money for such a small car, keep in mind that not only is the Smart very solidly built, but it also comes standard with side air bags, advanced EPS electronic stability program (which applies brakes to individual wheels to prevent swerving), and many other advanced features completely unheard of in this price class. The Canadian version of the Smart gets 51 mpg in the city, and 62 mpg on the highway, giving it the same average mileage as the Prius. However, keep in mind that diesel emits 15 percent more CO_2, unless you use biodiesel, in which case it emits 78 percent *less* CO_2 than gasoline.

Carbon Buster/Miser Recommendation: Replace secondary family car (24 mpg) with a Mercedes Smart (56 mpg).

5-year savings: $3,475, 24,400 lbs. CO_2, 1,246 gallons gasoline.

Life-time savings: $11,410, 80,000 lbs. CO_2.

Incremental cost: $750 premium for diesel engine. **New cost:** $14,275.

Payback: 1.1 years. **IRR:** 93%. **CROI:** 107 lbs./$. Note: available in Canada. US availability 2006 (Zapworld.com) or (Smart.com).

Efficient Gasoline-Powered Cars

What about the good old gas-powered car? Are there any that are even worth considering when shopping for an energy efficient vehicle? The bad news is that today's gas-powered cars have become *less* efficient than those of 15 years ago, both in terms of average performance and in terms of the most efficient car: For example, the Honda Civic HB VX, produced in 1992, and the Chevy Sprint (also known as the Pontiac Firefly, Geo Metro and Suzuki Swift) achieved 51 mpg and 45 mpg, respectively, while today's most efficient gas-powered car, the Toyota Yaris, gets only 38 mpg.

Today's best gas-powered cars are the Toyota Echo/Toyota Yaris (38 mpg), and the Toyota Corolla (36 mpg).

Carbon Miser Recommendation: Replace primary family car (assumed to have an efficiency of 17 mpg) with a 2005 Toyota Echo (or Toyota Yaris or Corolla).

5-year savings: $5,037, 35,300 lbs. CO_2, 1,805 gallons gasoline (less for Yaris/ Corolla).

Life-time savings: $15,990, 112,000 lbs. CO_2.

Incremental cost: $0. **New cost:** $10,985.

Payback: 0 years. **IRR:** ∞ percent. **CROI:** ∞ lbs./$. (Since the Echo has no incremental purchase cost, these numbers cannot be calculated)

A Word on Performance

My Webster's Dictionary defines performance as "the manner in which or the efficiency with which something reacts or fulfills its intended purpose."

> **Quick Tip**
>
> The Toyota Echo/Yaris is the Carbon Miser's top choice, due to its low cost and very low fuel consumption (Figure 6.8).

I particularly like the word "efficiency" in that definition. Yet, when it comes to common parlance about car "performance", the word "efficiency" has been almost entirely replaced by acceleration.

If you are a race car driver, or if your day-job involves making quick getaways from banks, large satchels of money gripped tightly under your arms, acceleration may be a very important component of performance to you. But, honestly, can you remember the last time you desperately needed to accelerate your car from 0 to 100 mph in 6.9 seconds or less? From 1981 to 2003, the average new car horsepower increased by 93 percent, while fuel consumption between 1987 and 2004 increased by 6 percent.[50] What this means is that we have managed to roughly double the efficiency of car engines in the last 25 years, but rather than translate this into cars that are more fuel efficient, we now have cars with oversized motors. This is one of the reasons why hybrids are so efficient. Regular cars have an engine that is sized for the occasional burst of speed. This oversizing makes the engine inherently inefficient, since 99 percent of the time it has to operate at less than its optimum.

Hybrid gasoline engines, on the other hand, are sized closer to typical running loads, with the electric motor, which can deliver extremely high torque, providing the extra acceleration when needed.

One of the reasons for the ever-increasing spiral of horsepower ratings of new cars may lie in the nature of many reviewers: naturally, car enthusiasts often fulfill this role, enthusiasts whose resumes may include stints at *Hot Rod* or *Circle Track*. In the 1950s, Detroit's high-performance cars had 150 horsepower engines.[51] Now, reviewers enthuse about 500 hp engines for passenger cars.

The question is, do these reviews serve the needs of average car buyers? The disparity is perhaps nowhere more apparent than when comparing editors' ratings and the ratings of car owners. For example, on one web site rating the (not overly high-powered) Honda Insight, professional reviewers rated it a full 33 percent lower than actual consumers, the greatest difference I have seen in any review.[52]

While the wish to pass a large semi trailer driving below average highway speeds is reasonable, a very strong engine may actually be deleterious to your health, a fact evinced by higher accident and car insurance rates for higher-powered cars.

Biofuels and Natural Gas
COMPRESSED NATURAL GAS (CNG)

Another approach to reducing car greenhouse gas emissions is through fuel switching. For example, by using natural gas instead of gasoline, the reduction of CO_2 emissions is about 34 percent. At the time of writing, natural gas is about 32 percent cheaper than gasoline, and the highest US federal tax credit, $4,000, is for the Honda Civic GX, which runs on compressed natural gas (CNG). If you live close enough to a natural gas refueling station and do not need to drive to

Carbon Miser: 20.9%; $8,512.00

remote locations, CNG can be cost-effective and environmentally beneficial, with a very good payback.

For some of the advantages and disadvantages of CNG conversion, check out the web site of The Energy and Resources Institute: Static.teriin.org/energy/cng.htm#adv

5-year savings: $3,374, 18,588 lbs. CO2.

Life-time savings: $11,077, 61,000 lbs. CO2.

New cost: $2,000. **Payback:** 3.0 years. **IRR:** 33.9 percent **CROI:** 30.5 lbs./$.

BIOFUELS

Natural gas, though lower in emissions, is still a fossil fuel, meaning that it contributes a large net increase to atmospheric carbon dioxide. On the other hand, biofuels, such as ethanol (alcohol) or biodiesel (for example processed vegetable oil) have the potential to emit little or no net CO2 emissions.

This is because the carbon dioxide released during combustion is equivalent to the carbon dioxide that was captured by the plants during photosynthesis. However, the carbon efficiency also depends on the energy required to grow and harvest the plants, and to process the fuel.

One researcher has pegged the energy return on investment (or EROI) of ethanol from corn, and corn residues, at 1.3, and 0.7 to 1.8, respectively (EROI is the amount of energy recovered divided by the amount of energy invested to obtain the energy).[53] This means that ethanol from corn residues *can* have a negative net energy gain, i.e. under certain conditions ethanol can take more energy to produce than it delivers.

With all forms of biomass energy, we have to be very careful to ensure that the processes used in their production are sustainable, i.e., have a positive carbon balance. For example, a homesteader using dead-fall or selective logging to heat his or her home may not add any net carbon to the atmosphere. On the other hand, if wood use were to become so intensive that it led to large-scale clear-cutting, the carbon balance would become negative. Similarly, a farmer using organic growing techniques to produce ethanol distillation feedstock, and using ethanol for farming machinery herself, could likely produce net-zero carbon fuel (or very close to it). However, ethanol produced through intensive application of fossil-fuel based fertilizers and pesticides could easily create an additional source of carbon emissions to the atmosphere.

BIODIESEL

A 1998 biodiesel lifecycle study, jointly sponsored by the US Department of Energy

Biodiesel works in any diesel engine with few or no modifications to the engine or the fuel system. Biodiesel has a solvent effect that may release deposits accumulated on tank walls and pipes from previous diesel fuel usage. The release of deposits may end up in fuel filters initially, so fuel filters should be checked more frequently at first.[55]

and the US Department of Agriculture, concluded biodiesel reduces net carbon dioxide emissions by 78 percent compared to petroleum diesel.[54]

According to the National Biodiesel Board:

The recent switch to low-sulfur diesel fuel has caused most Original Equipment Manufacturers (OEMs) to switch to components that are also suitable for use with biodiesel. In general, biodiesel used in pure form can soften and degrade certain types of elastomers and natural rubber compounds over time. Using high-percentage blends can impact fuel system components (primarily fuel hoses and fuel pump seals) that contain elastomer compounds incompatible with biodiesel, although the effect is lessened as the biodiesel blend level is decreased. Experience with B20 has found that no changes to gaskets or hoses are necessary.[56]

Many biodiesel enthusiasts are driving new, entirely unmodified, "off-the-rack" diesel cars with no apparent ill effects to the car engine.

Note: Just as with non-certified gasoline, using non-certified biodiesel may void your car's warranty. If you are worried about this, check with the manufacturer and/or your fuel retailer. Also, 100% biodiesel can gel at temperatures below 32° F.

 Carbon Buster Recommendation: Use biodiesel for Smart secondary family car (assuming 50 percent of diesel replaced with biodiesel).

5-year savings: $0, 21,400 lbs. CO2.

Incremental cost: $119 per year.

Payback: N/A, **IRR:** no personal financial gain, but large environmental benefit to the planet. **CROI:** 36.0 lbs./$.

Access:

US Biodiesel Association: National Biodiesel Board (NBB), biodiesel.org Canadian Renewable Fuels Association: greenfuels.org Where to buy biodiesel: US: biodiesel.org/buyingbiodiesel Canada: greyrocksmedia.com/ biodiesel/canadiannetwork/ availability.htm

The Wild, Weird and Wonderful

There are numerous other non-car motorized vehicles that may suit your particular needs as a car alternative for specific applications. Here is a selection:

 Carbon Miser: 20.9%; $8,512.00

- Segway: a gyroscopic, battery-powered, scooter-like robot (a "robotic mobility platform") that has an intuitive steering mechanism useful for downtown travel. The US Postal Service and Amazon.com are testing Segways' usefulness for postal delivery, and filling orders in their warehouses, respectively (Segway.com).

FINE MOBILE GmbH

6.11: *The TWIKE is a light-weight electric car with a top-speed of 53 mph, and the ability to switch to pedal power (some models). TWIKEs consume a mere 8 kWh per 100 miles and will cost you a mere 94 cents in electricity for this distance, compared to $13 in gasoline for the average 21 mpg car, i.e., 14 times less per mile.*

- Twike: a short-range, lightweight electric car that can also be propelled by human pedal power (Fig. 6.11). Starts at $20,385, depending on Euro exchange rate (Twike.com).

- Build your own electric car: If you can accept the idea of an electric car as your short-range secondary family car, good for 85 percent of your trips such as grocery shopping, picking up the kids or commuting — as opposed to cross-country tours — think about making your own, and save on gasoline. *Make* magazine (Makezine.com) has an article on how to get started ("Technology on your Time," Volume 05, page 60; Charles Platt. 2005. Electric Avenue) or subscribe and download the digital version at Makezine.com/05/electriccar.

- Motorcycles: possibly the world's only fuel economy guide to motorcycles: totalmotorcycle.com/MotorcycleFuelEconomyGuide/index.htm.

- Shell Eco-Marathon: these vehicles are built by high-school students to compete in Shell's annual Eco-Marathon. The aim of the race is to complete the race track with the least amount of fuel, at a minimum average speed of 19 mph. The vehicles, essentially glorified gas-engine powered three-wheeled bicycles with a carbon fiber body, are not for sale. However, the astounding fuel efficiency of these vehicles (the current record stands at 10,705 mpg) gives one the impression that modified versions of these vehicles, perhaps driving three abreast in special lanes, may eliminate much of the congestion and harmful emissions in our future cities. (At current fuel prices, and if you could drive on the equator, it would cost you $6.46 to circle the earth with one of these vehicles. Similarly, if you were to do all your driving in one of these, six dollars would be your approximate annual gasoline bill.)[57]

Carbon Buster: 32.6%; $10,197.70

- Electric bicycles: with over 15 million of these in use in China, electric bicycles can be purchased either as a complete package (about $800), or as a retrofit (about $300).

Access: Zapworld.com, Bionx.ca.

Car Safety

As Amory Lovins points out, there is no direct relationship between car weight and safety. Statistics compiled by Tom Wenzel, a scientist at Lawrence Berkeley National Laboratory, in California, and Marc Ross, a physicist at the University of Michigan, show that, for example, the nimble Jetta has less than half the fatalities (47 per million cars) of the bestselling vehicle in North America, the Ford F-series pickup truck (110 per million cars). The study was adjusted for the demographics of the car drivers, e.g. age and gender.

The picture looks even worse when comparing how many deaths of individuals not riding in the car are caused by the vehicles: the Ford F-series leads to the deaths of over five times as many people per million cars as the Jetta.

Other cars that do well in low fatality rates are the Honda Accord and Toyota Camry.

The study, which was commissioned by the US Department of Energy, is available at the website of the American Council for an Energy Efficient Economy: aceee.org/pubs/t021full.pdf.[58]

Electric Power

Your Home's Power Bill

E LECTRICITY IS THE SECOND BIGGEST ITEM on your family's energy bill, responsible for 34 percent of the carbon emissions and 16 percent of the cost (42% of your *utility* bill).

Electrical power consumption is billed in kilowatt-hours, or kWh. One kilowatt-hour is consumed when a 1,000-watt appliance, for example an electric heater, runs for one hour. Running a 100-watt incandescent light bulb for ten hours or a 25-watt compact fluorescent light (with the same light output of a 100-watt incandescent bulb) for 40 hours also consumes one kWh (Fig. 7.1).

A typical house consumes 11,782 kWh in one year, or 32 kWh per day.

Another way to get a feel for how much energy is in one kilowatt-hour (1 kWh) is to compare it to human energy: the average

"All power corrupts — but we need electricity."

— Anonymous

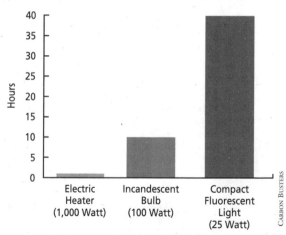

7.1: *How long can an appliance run on one kilowatt-hour?*

human consumes about 100 watts continuously, about as much as a typical light bulb, one kWh in 10 hours, or about 2.4 kWh per day (more during hard physical labor). This means that your home's average power consumption is equivalent to 13 people, working 24 hours around the clock. So, in effect, you are employing 13 "energy serfs." And you thought royalty had it good!

The current national average cost for power is 9.89¢ per kWh.[59] If you consider that one kWh is equivalent to the energy consumption of ten people working for 0.989 cents per hour each (i.e. less than a penny per hour), you get an idea of why we use power as thoughtlessly as we do: it's dirt cheap!

Yet, if you add up our national consumption and look at the environmental and monetary cost of electricity, the impact is quite significant. And despite

Kilowatt-hours and Kilowatts

What's the difference?

kWh

Energy consumption, especially electrical consumption, is measured in kilowatt-hours (kWh). KWhs indicate the cumulative power consumption over time. For example the use of 50 watts for 20 hours, or 1,000 watts for one hour, both result in the consumption of one kWh.

A typical 2,000 to 2,500 sq. ft. free-standing single-family house (that forms the basis of our calculations here) consumes 11,782 kWh per year. The current national average retail cost for one kWh of residential power consumed is 9.89 cents per kWh.

1,000 watt-hours = 1 kilowatt-hour (kWh)

1,000 kilowatt-hours (kWh) = 1 megawatt-hour (MWh)

1,000 megawatt-hours (MWh) = 1 gigawatt-hour (GWh)

1,000 gigawatt-hours (GWh) = 1 terawatt-hour (TWh)

So, for our typical house, the annual consumption is 11,782 kWh, or 11.8 MWh.

kW

Instantaneous consumption of electrical equipment is usually measured in watts or kilowatts (1,000 watts equal one kW). For example, a 60-watt light bulb will consume 60 watts continuously for as long as it is on. Other equipment, such as refrigerators, vary in power consumption depending on whether or not the compressor is running at the moment. The nameplate rating of electrical equipment therefore either indicates the *actual* consumption (e.g.: in the case of light bulbs) or the *maximum* consumption (e.g.: in the case of computers).

Individual homes are generally not billed for their maximum kW demand, but a kW charge often applies to larger buildings, particularly commercial, institutional and industrial facilities. While maximum kW consumed have no direct relationship to consumption, energy service providers have to provide power production capacity based on the annual peak power demand, which is considerably more expensive than supplying average consumption. The kW penalty is also known as a "demand" or "peak demand" charge.

Carbon Miser: 20.9%; $8,512.00

electricity's relatively low cost, there is still tremendous unrealized economic potential in energy conservation and energy efficiency, even at today's low prices.

Lighting

Saving lighting energy is one of the easiest, cheapest and most satisfying ways to reduce energy costs in your home (most satisfying because light is much more visible than other forms of power consumption). Lighting consumes an average of 10 percent of our home power (7 to 15 percent, depending on the study you look at). This can vary quite a bit, of course, depending on each home's lighting configuration. Because of its high visibility, light consumption tends to get overestimated by the average user. Still, it represents a considerable chunk of your power bill.

Incandescent Lights

These are the lights we commonly mean when we are talking about a light bulb. However, given their low efficiency, we can think of them more accurately as small electric heaters: 90 percent of electricity entering a light bulb is immediately converted to heat (Figure 7.2). Moreover, an incandescent bulb burns for only about 1,000 hours before it has to be replaced.

Incandescent lights are so inefficient that the last one was banished from my off-grid home 16 years ago. A few years later, when we were doing some efficiency testing comparing compact fluorescent lights (see below) and regular incandescent bulbs, we actually had to go out to purchase some incandescent lights, since not a single one could be found in the house! The only incandescent lights remaining in our solar home are the fridge light and flashlights. And even our flashlights are quickly being replaced by more efficient LED-powered models.

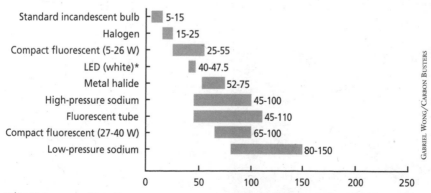

7.2: *Efficiency of selected lamp types in lumens per watt (higher numbers are better; sources: RMI E-Source, Philips, TCP).*[60]

Carbon Buster: 32.6%; $10,197.70

Although still widely available, incandescent bulbs are essentially obsolete now except for a few specialty applications. They are a 19th-century invention best left behind in the 20th century.

ENERGY-SAVING INCANDESCENTS

These are simply bulbs that have slightly lower-rated wattages than the standard bulbs, e.g.: 52 watts instead of a standard 60-watt bulb. They do not offer more light per watt than regular bulbs.

LONG-LIFE INCANDESCENTS

Special incandescent light bulbs are available that last up to 3,000 hours, instead of the usual 1,000. Longevity is one important part of the sustainability equation, as it means that fewer resources are consumed for a given amount of time. However, long-life bulbs are actually less efficient than regular bulbs.

Long-life incandescents achieve their longer life by either having a thicker tungsten filament, or by "underdriving" 130-volt bulbs at the rated 120 volts (nominal voltage 110 volts), reducing filament temperature, wattage and efficacy. Either way, you actually lose 20 to 30 percent of your light output per watt.

Therefore compact fluorescent lights are a better solution for home applications.

Compact Fluorescents (CFLs)

Compact fluorescent lights (CFLs) are the way to go for most residential lighting applications (Figures 7.3, 7.4).

7.3: *A triple biax compact fluorescent light (CFL).*

CFLs are four times as efficient as incandescent bulbs, representing a whopping 300 percent increase in efficiency. Also, they last 6 to 15 times longer, with typical lifespans of 10,000 hours. The rated life of a lamp is the point in time at which an average of 50 percent of lamps fail. Some will fail earlier, some later. CFLs need to be replaced much less frequently than incandescent lights, saving you both time and money.

7.4: *A spiral compact fluorescent light (CFL).*

When you buy CFLs, keep in mind that some models are larger than the

Carbon Miser: 20.9%; $8,512.00

incandescent bulbs they replace. Spiral CFLs greatly reduce size requirements (e.g.: the 42-watt GE T4 Spiral replaces a 150-watt incandescent and fits a harp as small as 8.5 inches).

CFL tubes contain small amounts of mercury and need to be disposed of as toxic waste, just like regular fluorescent tubes (however, the amount of mercury is far less than what would have been released from coal mining to supply the extra energy needed for incandescent light bulbs). Check if lamps can be recycled in your area, to further reduce environmental impact: lamprecycle.org. For example, Philips' ALTO line of lamps uses 100 percent recycled mercury.

> ## Quick Tip
> Compact fluorescent lights (CFLs) beat incandescent bulbs by a ratio of 4:1, and replacing a 100-watt incandescent bulb with a 23-watt CFL saves you $76 in electric power costs, plus 980 pounds of CO_2 over the life of the lamp.

Like most electronic equipment, such as TVs or computers, CFLs create an electromagnetic field. While the evidence is still inconclusive, there is some indication that these electromagnetic fields may be harmful to human health. It is a good idea to stay at least at arm's length from a CFL (as well as cathode-ray tube TV and computer monitors), at which distance the field is much reduced.

Warning: Regular CFLs *cannot* be used with dimmer switches! For CFLs suitable for dimmers, please see page 60.

The color "temperature" of a lamp (expressed in "K") indicates the relative "warmth" of the light to the human eye; e.g.: 2,600 K light is relatively warm and yellowish, while 4,000 K light is considered cold white, or bluish.

Many modern CFLs are so close in light quality to conventional incandescent light bulbs as to be virtually indistinguishable for all but the most discerning eye. However, you have to know which CFLs to buy, and which to avoid, if you are concerned about the color temperature. We have tested a variety of CFL brands and models, shown in Fig. 7.5. In areas where quick starting and light quality are of prime importance, we recommend the models marked with the Carbon Busters logo in Fig. 7.5. In areas where these features are less important, you can let price and other features be your primary guide.

HIGH-QUALITY ELECTRONIC BALLASTS

Lights equipped with these ballasts come on immediately, or nearly so. All electronic ballasts cycle at 10,000 hertz, so they never flicker, unlike older-style magnetic ballasts still found in many offices.

CHEAPER ELECTRONIC BALLASTS

Cheaper CFLs often also have cheaper electronic or even magnetic ballasts. CFLs powered by a cheaper ballast usually take a few seconds to "plink" on, just like

Carbon Buster rating[1]	Brand	Approximate cost ($)	Life in hours[1]	s-spiral t-tubular g-globe a-a-type, f-f-flood c-chandelier	Nominal wattage output	Measured wattage[2]	Nominal lumen	K (color temp.)	Seconds to light	Energy Star Rated
5	Philips Marathon 11 W	7	8,000	s, n	11	8.3	675	2,700	instant	yes
5	Philips Marathon 23 W	5	6,000	s, n	23	17.9	1400	2,700	instant	yes
5	General Electric soft white 120	11	6,000	s, n	32	29	2150	-	0.6	no
5	TCP T2 Springlamp - 9 W	7	10,000	s, n	9	6.3	550	2,700	instant	no
5	Sylvania Dulux	3	10,000	t, n	23	21.7	1247	3,000	1.3	yes
4	Globe Ener Saver Mini Cool White	7	6,000	s, n	9		500	4,100	1.3	no
4	Sylvania Soft White 60	7	6,000	a, n	14	10.9	800	3,000	1.0	yes
4	Rona Fluorescent mini spiral	4	6,000	s, n	13	9.1	800	3,000	instant	no
4	Sylvania Soft White mini 60	6	8,000	s, n	13	10.4	800	3,000	instant	yes
4	Globe Ener Saver Soft White	7	6,000	s, n	9	6.1	500	2,700	0.9	yes
4	TCP T2 Springlamp - 13 W	11	10,000	s, n	13	10.3	900	2,700	instant	yes
3	General Electric soft white 40	7	6,000	g, n	9	6	430	-	1.7	yes
3	Noma G25 Soft White Globe Bulb	6	6,000	g, n	9	6.3	475	2,700	instant	yes
3	Sylvania Daylight Extra mini 60	6	8,000	s, n	13	10.4	800	3,500	1.2	yes
3	General Electric soft white 60	9	8,000	s, n	13	9.3	825	-	0.7	yes
3	Sylvania soft white mini 100	7	8,000	s, n	23	18.4	1600	3,000	1.0	yes
3	General Electric soft white 75	-	6,000	t, n	20	16	1200	-	instant	yes
3	General Electric soft white 60	7	8,000	a, n	15	11	850	-	1.0	yes
3	General Electric soft white 100	11	8,000	s, n	26	20.6	1700	-	1.2	yes
3	General Electric soft white 50, 100, 150	13	6,000	s, d	12, 23, 32	18.4	600/1600/2150	-	1.0	no
3	Sylvania soft white 100	9	10,000	s, n	27	21.8	1750	3,000	0.9	yes

Carbon Miser: 20.9%; $8,512.00

Carbon Buster rating[1]	Brand	Approximate cost ($)	Life in hours	s-spiral t-tubular g-globe a-a-type, f-flood c-chandelier	Nominal wattage output	Measured wattage[2]	Nominal lumen	K (color temp.)	Seconds to light	Energy Star Rated
3	IKEA 20W		10,000	t	20	19		-	4.0	no
3	IKEA 9 W globe		n/a	g	9	9.1		3,300	2.9	no
3	Sylvania soft white Globe 60	-	6,000	g, n	15	13.2	700	3,000	instant	no
3	IKEA 11 W	6	6,000	t, n	11	9.9	600	-	instant	no
2	Globe Ener Saver chandelier	8	6,000	t, n	7	5.8	330	2,700	1.9	no
2	IKEA 15 W	6	10,000	t, n	15	14.2	950	-	1.0	no
2	TCP T2 Springlamp - 2 W	7	8,000	s, n	2	2.1	125	2,700	instant	no
1	General Electric soft white 40	6	6,000	a, n	11	8.4	520	-	1.1	yes
1	General Electric soft white 50	8	6,000	f, n	11	8	380	-	0.7	yes
1	Philips Daylight 15 W	10	8,000	s, n	15	11.7	950	-	instant	no
1	IKEA 4 W	13	10,000	g, n	4		127	-	2.2	no
1	IKEA 7 W	13	10,000	g	7			-	1.0	
-	General Electric Floodlight 90	10	6,000	f, n	26		1200	-		yes
-	General Electric soft white 25 chandelier	7	6,000	t, n	7	6.3	370	-	0.5	no
-	Sylvania Flood 65		6,000	f, n	15	12.3	560	3,000	<1.0	yes
-	IKEA 10 W chandelier		n/a	c	10		500	-	1.0	
-	Purlite Pure natural soft 15 W		n/a	s, n	15			-	instant	
-	Satco 23 W dimmable		n/a	s, d	23			3,000	instant	

7.5: *Characteristics of select compact fluorescent lights (CFLs). Top choices are marked with the carbon buster's logo (highest rating 5, lowest rating 1. "-" indicates data not available. Source: Claudia Bolli, Michèle Elsen/Carbon Busters).*

Code: s = spiral, t = tubular g = globe a = a-type, f = flood c = chandelier, d = dimmable, n = non-dimmable; 1 = based on subjective light quality, and time to start/reach full brightness; 2 = measured 4 minutes after lamp turned on.

Carbon Buster: 32.6%; $10,197.70

older fluorescent lights in offices. Also, they may take a considerable amount of time before they reach their full light output.

CFLs with cheaper ballasts are most suitable for areas that are lit for a long time, or where an "immediate-on" light is not required. They are less suitable for areas where you need immediate light for a brief time, such as hallways.

Plug-In Lamps (Two-Piece CFLs)

You can also get two-piece CFLs that feature separate ballasts and lamps. When the lamp is burned out, e.g.: after 10,000 hours, you can re-use the ballast for three to four more lamps (total ballast life of 40,000 to 50,000 hours). The advantage of plug-in CFLs is that you will save even more money, since replacement bulbs are generally cheaper than replacing the ballast with a one-piece CFL.

Also, by using two-piece CFLs, you are eliminating the energy and material costs of an additional three to four ballasts for one-piece CFLs, with the concurrent environmental benefits. However, virtually all new CFLs are provided as one-piece lamps, including the ballast.

Outdoor CFLs

Just like regular fluorescent tubes, CFLs don't start or run well at very low temperatures. If a CFL is in an enclosure, it will usually supply enough heat to run well, but starting may still be a problem.

If you live in an area where temperatures drop to below 14°F, you may want to purchase cold-starting CFLs for all outdoor applications. Examples include the GE Reflector 15-watt 49917 rated to -22°F and the Philips Marathon Universal, Outdoor and Flood (-10°F).

Dimmable CFLs

Regular CFLs must never be used in dimmable sockets. Special dimmable CFLs are available. Dimmable CFLs include all of GE's 4-Pin Double and Triple Biax lamps, some of GE's 2-Pin Biax lamps, and Philips Marathon Dimmable 23-watt CFLs. (You can also replace your dimmable wall switch with a regular wall switch, and then use regular CFLs.)

3-Way CFLs

There are also CFLs that replace 3-way switchable incandescent bulbs (usually 150-watt incandescents). GE Biax 29-watt 41327 and 41442 CFL, and GE Circlite 2D-Electronic 39-watt 25809/25812, and 27253, Philips Marathon 3-way 34-watt bulbs fall into this category.

Full-Spectrum CFLs

Full-spectrum light is generally considered the second healthiest lighting choice (after daylight). Full-spectrum CFLs were pioneered by the OTT-Lights Ottlite.com, and are also available from other manufacturers. Keep in mind, though, that sunlight is actually rather white in appearance, so you may find full-spectrum lights a little colder in feel than what you are used to. Also, fluorescent light cannot truly represent the color spectrum of the sun smoothly, as each of the gases in the mini-tube tends to emit at discrete portions of the light spectrum. Nevertheless, one study found that full-spectrum fluorescent lighting, not blocked by plastic covers, and with specular reflectors without translucent coatings, reduced caries (tooth decay) in school children in Alberta by 27 percent, apparently due to the UV-A component of the lights.[61]

Carbon Buster/Miser Recommendation: Replace incandescent bulbs with Compact Fluorescent Lights (CFLs).

5-year savings: $350, 4,500 lbs. CO_2, 3,500 kWh of power.

Life-time savings: $999, 12,800 lbs. CO_2 at an average lamp life of 14 years, or 10,000 hours of use.

Incremental cost: $117. Based on 13 high-quality CFLs for $7 each, five inexpensive CFLs at $2 each for less important areas, and two CFLs at $17 each for specialized uses, such as dimmability, 3-way switching or outdoor use. You will save $18 by not having to buy 72 replacement incandescent bulbs at 25¢ each over the next five years. It is assumed that 20 percent of your lights are already CFLs, or cannot be easily retrofitted with CFLs. **New cost:** $135.

Payback: 1.7 years. **IRR:** 60 %. **CROI:** 110 lbs./$.

Halogen Lights

Halogen lights are actually a specialized form of incandescent bulb: they use halogen gases to redeposit vaporized tungsten on the filament, thereby preventing darkening of the bulb. Halogen lights are about 40 percent more efficient than regular incandescent bulbs, but not nearly as efficient as CFLs.

HALOGEN DESK LAMPS

IKEA has popularized the use of the 20-watt halogen desk lamp; it is fairly cheap, small and reasonably energy efficient, though less than half as efficient as a compact fluorescent lamp.

One thing to watch out for: the power adapter (black power brick) draws power even when the lamp is not in use. In fact, even though the halogen bulb is rated at 20 watts, the total lamp consumption is 25 watts, since the power brick draws 5 watts. This means that over a year, the lamp will consume more power while it is off than while it is on!

Recommendation: Unplug it when not in use, or switch it on and off with a power bar, or switch to compact fluorescents. (See p. 67 for more information on power vampires.)

Carbon Buster: 34.1%; $10,547.30

TORCHIÈRES: CARBON MONSTERS

About ten years ago, torchières were very popular. Torchières usually have a floor-based lamp stand topped by a concave bowl which contains the halogen bulb. They were so cheap (often $10 to $15) that students especially bought them in large numbers. The light points upwards towards the ceiling, creating indirect lighting. Unlike the desk-lamp version, though, they have bulbs that use up to 500 watts. 500 watts! That's enough to power 20 to 50 compact fluorescent lights.

Torchières are hugely inefficient. Worse, they represent a fire hazard, since paper left on top of the lamp, or even pieces of dust on top of the quartz glass protecting the bulb, can cause a fire. Also, the bulb emits UV light that can damage the eyes.

Modern versions of these lamps have been limited to a maximum of 300 watts to reduce the risk of fire, and generally have a UV-absorbing cover. Yet even the new versions are still huge energy wasters and carbon emission monsters.

Recommendation: Check the watt rating of your torchière. If it consumes more than 100 watts, terminate its existence swiftly and painlessly.

Fluorescent Lights

T-8 AND T-5 TUBES

The fluorescent tubes commonly found in offices and schools are just as efficient as compact fluorescent lights (see "Compact Fluorescents", p. 56), in fact, even slightly more efficient. However, their light is usually too bright for most home uses, so residential applications are normally restricted to the kitchen, and sometimes the bathroom, laundry room or basement.

If you already use fluorescent tubes, you can replace older T-12 tubes with more efficient T-8 or even T-5 models, and upgrade to a hum- and flicker-free, more energy-efficient electronic ballast. The numeral in the tube's designation refers to eighths of an inch. Therefore, a T-12 has a tube diameter of 1.5 inches, a T-8 is 1 inch in diameter, and so forth. This means that you can easily determine your type of tube with a ruler.

FULL-SPECTRUM FLUORESCENTS

Full-spectrum versions are also available for fluorescent tubes. They are actually slightly less efficient (and a lot more expensive), but provide healthier light.

LEDs

The latest lighting development is in the area of LED lights (Light Emitting Diodes). In particular, the development of white LEDs has moved them into the realm of residential use.

LEDs are most suitable as point sources, since they emit usable light even at a fraction of a watt. By contrast, incandescent bulbs and CFLs operate at greatly

 Carbon Miser: 22.5%; $8,861.60

Common Fluorescent Tube Myths

"It takes more energy to turn a fluorescent tube on than is saved by switching it off."

It is true that fluorescent tubes consume slightly more energy during the start-up phase to establish the initial start-up voltage for ionizing the mercury vapor and argon gas. However, this spike lasts only $\frac{1}{60}$th of a second, and is not even registered by the power meter.

"It is better to leave fluorescent tubes on, since switching burns out the tubes."

Unlike incandescent bulbs, each ignition does indeed wear out the cathode a little bit. In the early days of fluorescent tubes (1950s) this effect was so pronounced that tubes were left on 24 hours per day to avoid the "loss-of-life" penalty for switching them on and off. However, since then technology has advanced considerably, plus the cost of tubes has dropped. Now, a tube may lose only ten minutes of total life when switching it on and off. Keep in mind, though, that if you leave a tube on for ten minutes while you are away, you are also losing ten minutes of tube life. Plus, you have to pay for the power in addition to the tube depreciation. In fact, over the life of the tube, you will pay about 29 times more for the electricity consumed to power the tube than for the tube itself. This means that today it is cost-effective to switch off the tube even when leaving the room for as little as one or two minutes.

reduced efficiency at lower wattages, or are not even available at very low wattages in the case of CFLs. Furthermore, LEDs are extremely shatterproof, work very well even under very low temperatures, and the bulbs can last from 50,000 to 100,000 hours, i.e., about ten times as long as CFLs and 100 times as long as incandescent bulbs. At these lifespans, the environmental and monetary savings for the displaced lamps alone can pay for the cost of the LED lamp.

LED lamps are about four times as efficient as incandescent bulbs; i.e., similar in efficiency to CFLs.

So, what's not to like? For one, the white light of LEDs is very white indeed, with a hint of blue, giving LEDs a cold feel that most individuals will not like for room lighting (though undoubtedly a few may actually prefer it). Also, LEDs are only available in lower wattages/lumen outputs at this point, though the availability of higher-wattage LEDs is increasing almost daily (currently, most LED lights range in milliwatts to about five watts in power consumption). Furthermore, the dispersion (distribution of light) tends to be poorer in LEDs, making them ideal for flashlights, but less ideal for room lighting (though again, advances in this area are made almost daily).

Carbon Buster: 34.1%; $10,547.30

LEDs are most suitable for outdoor lighting, spot lighting, Christmas lights (page 95), and night lights. A few intrepid souls are also using them as desk lamps.

Task Lighting

One trick for greatly reducing lighting power consumption is the use of task lighting. For example, if you are working at a desk, or reading a book, you may not need to illuminate the whole room. Therefore, a 15-watt CFL desk lamp would be more efficient than a 25 watt CFL room light, resulting in 40 percent power savings.

Motion Sensors

The only computer that can be mass-produced by unskilled labor is still the human brain. In a similar vein, an active human brain still outperforms any automated system for energy conservation in almost all applications. This also holds true for motion-sensor activated lighting. However, there are a few applications where motion sensors work very well.

One perfect application for motion sensors is outdoor lighting: here, a motion sensor dramatically reduces power consumption. Motion sensors also work well in areas where you may have your hands full, for example in entryways or entrance hallways. You can usually adjust the amount of time you would like the light to stay on after activation (e.g.: from one to ten minutes), and regulate the sensitivity (range), so that the sensor isn't activated by every pedestrian passing on the sidewalk.

Many outdoor motion sensor devices are also equipped with photocells, to prevent them from being activated during daylight hours (see "Photocells," below).

My experiences with cheap motion sensors bought in big-box stores have not been very good; they tend to fail after a short period of time. You are probably better off getting one from a specialty lighting store, often with a good warranty. The best combinations for outdoor use are motion-sensor-powered floodlights with LEDs, followed by CFLs or halogen lights (the CFL's life will be reduced somewhat through frequent switching, but it will still need to be replaced less frequently than an incandescent bulb). Even incandescent bulbs are not too bad in this application, since the light is "on" very little.

Photocells

Another way to control outdoor lights is through photocells. This works if you want the outdoor light to stay on all night. However, savings are lower than with motion sensors, and you will also create light pollution (see next section). Moreover, lights that are left on continuously are less effective at providing security and curbing vandalism than lights activated by motion sensors.

Carbon Miser: 22.5%; $8,861.60

Light Pollution

Light pollution is the obscuring of the natural night sky by unprotected sources of artificial lighting.

Light pollution is a bane particularly for stargazing astronomers: while excess light has little effect on the observation of the bright planets, it makes observation of deep space objects from the city nearly impossible.

So inured have we become to the starless night skies in our cities that when Los Angeles experienced a power blackout, the city observatory was inundated with calls from concerned citizens. The callers reported the sighting of strange objects and patterns in the night sky — their first view of the Milky Way!

Light pollution wastes a lot of energy by shining light where it has no business being; forward-thinking cities have started campaigns to reduce light pollution by installing street lighting that is focused tightly downward. Not only does this reduce glare for pedestrians and drivers, but the cities are realizing hefty savings in energy bills due to less light energy being sent into space.

Lights left on in office towers are also deadly for birds, and nature conservation organizations are urging businesses to ensure that lights are turned off at night.

The US Green Building Council recognizes light pollution as a serious problem, and even provides LEED rating credits to new buildings that do not emit light past property boundaries (LEED stands for "Leadership in Energy and Environmental Design). The LEED rating system is increasingly being adopted by many levels of government throughout North America as a minimum building standard for new government buildings. See usgbc.org for more information on the LEED system.

You can do your bit in the prevention of light pollution by designing your outdoor lighting so that it is not directed at neighboring structures, and only turns on when required.

Solar Tubes

Solar tubes are devices that conduct daylight into rooms, without the heat loss or gain usually associated with skylights. Solar tubes consist of a clear, light-collecting dome with a reflective backing, a long tube, usually 4 to 8 feet in length, that is also reflective, and a diffuser at the bottom. The reason that a solar tube is able to supply so much light is that, on average, horizontal surfaces receive 15 to 60 times the amount of light of vertical surfaces.[62]

A solar tube will cost you around $300.

 Carbon Buster Recommendation: Install a solar tube in an area that currently receives little daylight, but is frequently used during the day.

5-year savings: $108, 1,400 lbs. CO2, 1,100 kWh of power.

Life-time savings: $1,733 , 22,200 lbs. CO2, assuming life of 80 years.

Carbon Buster: 34.6%; $10,655.60

Incremental cost: $407, based on a purchase cost of $300, $110 dollars for your install time at $10 per hour, minus ten light bulbs saved over five years. New cost: $410.

Payback: 18.8 years. IRR: 5.2 %. CROI: 54.6 lbs./$.

Outdoor Lighting

The best lighting for the out-of-doors is motion-sensor controlled lighting, which reduces power consumption and light pollution (see "Motion Sensors," page 64, and "Light Pollution," page 65). The next best choices are LED lights (page 62) and CFLs (page 56 — low-temperature versions if you live in northern North America).

For yard lighting, metal halide, high-pressure sodium, and low-pressure sodium lamps provide even more output per watt than the other choices (Fig. 7.2). Disadvantages of these High-Intensity Discharge (HID) lamps are:

- Poor color rendering (usually yellow to orange)
- Not available in low wattages (only useful if you want to light large areas)
- Require a 15-minute cool-down period (restrike time) before reactivation can take place (i.e., not suitable for motion sensors).

So, HID lamps, while more energy-efficient, are only suitable if you have to illuminate large areas for long periods of time, and if color-rendering is not an issue.

Solar Lights

Solar lights are devices that combine a photovoltaic (electricity-producing) solar panel, a rechargeable battery and a (usually energy-efficient) light. There are two common applications of these lights:

MOTION-SENSOR FLOODLIGHTS

The main advantage of solar floodlights is that they do not require any wiring, and the solar panel can be placed more than ten feet from the battery, so that it can collect sunlight at an advantageous location and orientation. Of the two solar floodlights I have tested, one had an unreliable motion sensor. More recent versions contain LEDs, which make them quite effective.

GARDEN ACCENT LIGHTS

Solar accent lights for the garden do not provide enough light for reading — their purpose is purely decorative. Now that the solar accent lights run on LEDs, they last quite a long time, even with the Nickel-Cadmium batteries which they usually contain, as opposed to Nickel-Metal Hydride batteries, which have higher power density. Just like the solar floodlights, they contain photocells, so that they only come on at night, and turn off at dawn.

 Carbon Miser: 22.5%; $8,861.60

Solar-powered lights work well south of the 40th parallel. If you live further north, they will work well in the summer.

Solar floodlights with incandescent bulbs will draw too much power in winter further north. It remains to be seen how well new LED floodlights will fare. Really far north (e.g.: Alaska), solar lights are not much use: summer days are too long, and winter days too short.

LED-powered accent lights work well for much of the year, even as far north as the 55th parallel, but will not work properly from November to February.

Most solar accent lights have horizontal solar cells. If you live in the northern US or Canada, purchase those whose cells are nearly vertical for better winter use.

Power Vampires

What is a power vampire? Power vampires (or power leeches) are electrical devices that continuously draw power from your power outlets, even when not supplying any useful service. One example that we have encountered already is the power adapter for halogen lights. However, a modern home may easily have 10 to 30 power vampires: television sets, cable boxes, modems, satellite receivers, VCRs, DVD players, DVD recorders, fax and answering machines, computers, printers, copiers, wireless and cell phone chargers, wired and wireless hubs and routers, monitors, battery-powered power tools, stereos, boom boxes, shredders, speaker systems, iPod chargers, video camera chargers, night lights, plugged-in electric toys and game machines are among the most common power vampires. As a simple rule, the more gadgets in your home, the more power vampires. Is the power consumption of power vampires significant?

Power vampires usually consume only a few watts. However, consider that these devices consume power around the clock, 8,760 hours per year, and you will appreciate their impact; a recent study found that together they accounted for 5 to 20 percent of total home power consumption, even exceeding the traditionally highest user (the fridge) in some homes. Power vampires are the fastest-growing power users in our residences.

So, what are annual power vampire figures? Roughly speaking, each watt of vampire power costs you one dollar per year. So, if you have 25 power vampires consuming an average of 7 watts each, they will cost you $175 per year, and emit about 2,000 pounds of CO_2.

Garlic won't do much good against power vampires (though it *will* probably improve your health), but there are a number of ways to eliminate virtually all vampires effectively.

The simplest way to get rid of these power-suckers is to unplug them when not in use. An added advantage is that this measure is absolutely free.

Quick Tip

Power vampires can suck up to 20 percent of your power. Exorcising them will reduce your energy bill by $50 to $200 per year ($250 to $1,000 in five years).

Carbon Buster: 34.6%; $10,655.60

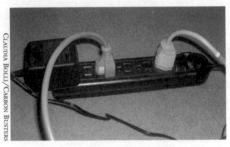

CLAUDIA BOLLI/CARBON BUSTERS

7.6: *Use a power bar to banish power vampires from your home.*

More convenient are power bars (Fig. 7.6). For about $5, you can lay one or several power vampires to rest.

You can extend the life of battery-powered equipment by not leaving it plugged in at all times, nor letting it run to zero for long periods of time. For example, a study of Apple laptop batteries found, curiously enough, that batteries in a charge state of 40 percent actually had the longest lifespan. Rarely used equipment (e.g.: battery-powered power tools) is best put on a timer-controlled power bar, to be charged for a few hours once a month or week; this way you reduce the power cost *as well as* extending useful battery life. Even if you double the life of a single rechargeable battery, you will probably have already paid for the timer three times over, without counting the power savings.

How can you identify power vampires in your home? There are a few simple rules:

If your equipment uses an external power adapter/power brick (usually black in color), it is *always* a power vampire.

If the equipment feels warm even when it has been switched off for a while, it is a power vampire.

If the power adapter is internal (i.e., no black external power brick), it could go either way:

Always a power vampire:

Laptop chargers	DVD players
Fax machines	DVD recorders
Answering machines	Digital video recorders
Printer	Photocopier with paper sorter
TVs	Phone chargers
Cable boxes	Hubs and routers
Cable modems	iPod chargers
Modems	Video camera battery charger
VCRs	Night light

Usually a power vampire:

LCD monitor	
Audio system	Photocopier without paper sorter
	Plugged-in electric toys

Carbon Miser: 22.5%; $8,861.60

May or may not be a power vampire:

Desktop computer	Boom box
CRT monitor	Battery chargers for rechargeable devices

Battery charger for AA, AAA, D, C, and 9V batteries (Nickel-Cadmium and Nickel-Metal hydride)

Never a power vampire:[63]

Incandescent lamps	Compact fluorescent lights

Exceptions to power-bar use for power vampires:

VCRs and DVD-Rs	Some computer servers, routers and hubs
Fax machines	Plug-in electric clocks

> **Carbon Buster/Miser Recommendation:** Eliminate 90 percent of your power vampires.
> **5-year savings:** $524, 6,700 lbs. CO_2, 5,300 kWh of power saved.
> **Life-time savings:** $2,622 , 33,700 lbs. CO_2, assuming a life of 25 years for the power bars.
> **New cost:** $125. $120 for 15 power bars at $5 each, three timers at $15 each, plus $5 for half an hour of your time to set up the power bars and program the timers at $10 per hour.
> **Payback:** 1.2 years. **IRR:** 84 %. **CROI:** 269 lbs./$.

Measuring Power Consumption

If you really want to know how much power your device draws in the "off" state, you need specialized equipment. The most accurate is a clamp-on amp-meter with a line splitter. However, at prices starting at around $100, this may be overkill. Also, you will need to measure amperage and voltage separately, and amp-meters will only give you instantaneous readings, making them nearly useless for equipment that cycles on and off (e.g.: refrigerators and freezers), and less useful for equipment whose power consumption changes significantly over time (e.g.: laptop and desktop computers, copiers, printers and fax machines).

More convenient are meters designed for consumers, for example the Watts up? and Kill-a-Watt power-consumption meters (Figures 7.7, 7.8). These devices can not only track average or total power consumption of an electrical device over time, but also automatically calculate the associated cost, once you have entered your utility rates. The only caveat is that for equipment using less than 5 watts or more than 3,000, the reading may be inaccurate or unavailable, since these meters are rated to measure consumption between 5 and 3,000 watts only.

Access: Kill-a-Watt power meter, $21.75
Manufacturer: p3international.com/products
Retail: ambientweather.com
Watts up? meter, $109.99 to 149.95. smarthome.com

RINA CHAN/CARBON BUSTERS

7.7: Two common tools for measuring home power consumption: the La Cross Technology Cost Control (left) and Watts-up? power consumption meters (right).

RINA CHAN/CARBON BUSTERS

7.8: Using a power meter to check for vampires.

Refrigerators
The Ten Most Efficient Fridges

The 10-cubic-foot Sun Frost RF-12 has been the leader in energy efficiency for many years, and still leads the pack with a mere 171 kWh in power consumption per year (Fig. 7.9). On the downside, Sun Frost fridges are hand-assembled for a small market (mostly, off-grid solar homes) and accordingly expensive: around $1,900.[64]

If your refrigeration space needs are modest, there are a number of fridges that cost a lot less, for example the no-freezer models from Avanti and BSH Continental, or the refrigerator/freezer combos from Classic 50's, Danby, and Summit, though none of these have quite the low power consumption of the Sun Frost RF-12, or the space of the Sun Frost R-19.

And if your fridge space needs are really, really modest, or you live in a small apartment, you can also consider a very small fridge (Fig. 7.10). Ironically, these small fridges actually have a fairly high kWh consumption per cubic foot. Nevertheless, due to their small size, the annual consumption is still low.

The Most Efficient Refrigerators by Size

If you need, or want, a larger fridge, check out Figure 7.11 (fridges over 10 cubic feet) and Figure 7.12 (fridges over 20 cubic feet). Or, check Figure 7.13 for the most efficient fridges on the basis of annual kWh consumption per cubic foot.

Of course, the power consumption does not tell you anything about features, style or manufacturing quality. You may want to check a recent copy of

Carbon Miser: 24.9%; $9,386.00

Brand	Model	Configuration	Volume cu. ft.	kWh/ year	kWh/ ad cf	Cost
Sun Frost	RF-12	Top Freezer	10.12	171	15.0	$1820-2059
Sun Frost	R-19	Refrigerator Only - Single Door	16.14	204	12.6	$2269-2539
Avanti	RM901W	Refrigerator Only - Single Door	8.70	230	26.4	$249-349
Avanti	BCA902W	Refrigerator Only - Single Door	8.87	247	27.8	$292-299
Sun Frost	RF-16	Top Freezer	14.31	254	15.1	-
BSH Continental	RC29	Refrigerator Only - Single Door	9.98	274	25.8	-
Avanti	1201W-1	Refrigerator/Freezer - Single Door	11.00	277	23.7	-
Classic	50's CBC960, CBC961	Top Freezer	9.50	285	28.7	-
Danby	D9501S, D9504W	Top Freezer	9.50	285	28.7	-
Summit	CM-115	Top Freezer	9.50	285	28.7	-

7.9: *The most efficient Energy Star fridges larger than 8 cu. ft. (based on EPA 2006 data and Internet research for retail pricing, adjusted volumes, ad cf = adjusted cubic feet, - = model discontinued or data unavailable).*

Brand	Model	Configuration	Volume	Adjusted Volume	kWh/ year	kWh/ ad cf
MicroFridge	MHRA-4E	Refrigerator/Freezer - Single Door	4.00	4.32	241	56
Whirlpool	EL5WTRXM*	Refrigerator Only - Single Door	4.00	4.00	250	63
Whirlpool	EL02CCXM*, EL02CCXR*	Refrigerator Only - Single Door	1.60	1.60	253	158
Whirlpool	EL02PPXM*	Refrigerator Only - Single Door	1.60	1.60	253	158
Avanti	651WT-1	Top Freezer	6.30	7.47	254	34
Samsung	SKR-A0742BU, SKR-A0752BU, SKRA0742BU, SKRA0752BU	Refrigerator Only - Single Door	2.47	2.47	256	104
MicroFridge	MHRB-4E	Refrigerator/Freezer - Single Door	4.00	4.32	259	60
MicroFridge	MHR-2.7E	Refrigerator Only - Single Door	2.70	2.70	260	96
Whirlpool	EL03CCXM*, EL03CCXR*, EL03PPXM*	Refrigerator Only - Single Door	2.70	2.70	262	97
Absocold	GARD562MG10R/L	Refrigerator Only - Single Door	5.60	5.60	268	48

7.10 : *The most efficient Energy Star fridges smaller than 8 cu. ft. (based on EPA 2006 data and Internet research for retail pricing, adjusted volumes).*

Consumer Reports for feedback on manufacturer's repair records and useful features. Unfortunately, *Consumer Reports* does not place a great emphasis on energy efficiency or environmental impact of the products it reviews. Still, it is a good starting point for researching product features and, to a lesser extent, longevity.

Carbon Buster: 37.0%; $11,179.90

Brand 10-20 cu. ft.	Model	Configuration	Volume	Adjusted Volume	kWh/ year	kWh/ ad cf
Sun Frost	R-19	Refrigerator Only - Single Door	16.14	16.14	204	12.6
Sun Frost	RF-12	Top Freezer	10.12	11.41	171	15.0
Sun Frost	RF-16	Top Freezer	14.31	16.77	254	15.1
Kenmore	7490*40*, 7491*40*	Top Freezer	18.79	21.94	387	17.6
Kenmore	7390*30*, 7393*30*, 7498*40*, 7499*40*, 7592*40*, 7594*40*,	Top Freezer	18.83	21.98	392	17.8
Kenmore	6397*30*, 7398*30*, 7397*30*, 6398*30*	Top Freezer	18.79	21.89	392	17.9
Kenmore	7395#, 7396#, 7595#	Top Freezer	19.00	22.35	405	18.1
LG Electronics	LRBC(N)20530** LRDC(N)20731**	Bottom Freezer	19.71	23.65	440	18.6
Maytag	MTB1953HE*	Top Freezer	18.50	21.94	413	18.8
Kitchen Aid	KTRP19KR**0*	Top Freezer	18.90	22.00	416	18.9
Whirlpool	ET9AHT*M*0*, ET9FHT*M*0*, GR9FHM*P*0*, GR9FHK*P*0*, GR9SHK*M*0*,	Top Freezer	18.88	22.05	417	18.9
Kenmore	6493*40*, 6494*40*, 6495*40*, 6496*40*, 7493*40*, 7494*40*, 7693*40*, 7694*40*	Top Freezer	18.87	22.02	417	18.9
Kitchen Aid	KTRC19KM**0*	Top Freezer	18.85	22.00	417	19.0

7.11: *The most efficient Energy Star fridges from 10.1 to 20 cu. ft. (sorted by consumption per cu. ft.). Based on EPA 2006 data and Internet research for retail pricing, efficiency based on adjusted volumes.*

All of the fridges listed in this book are Energy Star-rated by the EPA, which means that they are at least 15 percent more energy efficient than required by current federal standards and 40 percent more efficient than the conventional models sold in 2001. For updates on the latest Energy Star models go to:

Energy Star Access: energystar.gov.

Keep in mind that by the time the printed word reaches the bookstore, as with many consumer products, many models will already be discontinued. Still, the turnover rate is not nearly as high as in the computer world, updates are often minor, and the tables on these pages will provide you with a good benchmark. Unlike the automotive world, where average fuel efficiency has stagnated in the last 15 years, fridges are improving dramatically, so don't settle for less energy efficiency than what you see here.

What is the payback for replacing your fridge with a new energy efficient model? (See "What is payback?" p. 76.) Figure 7.14 lists average refrigerator efficiencies back to 1972. See if you can find the year your fridge was made from the

Carbon Miser: 24.9%; $9,386.00

Brand 20-31 cu. ft.	Model	Configuration	Volume	Adjusted Volume	kWh/ year	kWh/ ad cf
Monogram	ZIS480NR, ZISS480NRS	Side-by-Side	30.61	38.70	592	15.3
Kitchen Aid	KSSC48FM*0*, KSSO48FM*0*, KSSS48FM*0*	Side-by-Side	29.92	36.93	585	15.8
Kenmore	7420*40*, 7421*40*	Top Freezer	21.64	25.72	417	16.2
Profile	PTS25LBS, PTS25LHS, PTS25SHS, PTS25LHR*, PTS25SHR*	Top Freezer	24.60	29.10	475	16.3
Kenmore	7428*40*, 7429*40*, 7522*40*, 7524*40*, 7320*30*, 7323*30*	Top Freezer	21.64	25.73	422	16.4
Kenmore	7554#, 7555#	Bottom Freezer	25.00	29.6	499	16.9
LG Electronics	GR-B258****, GR-F258****, LRFC2575#**, LRFD2585#**	Bottom Freezer	25.00	29.6	499	16.9
Amana	ABB2524DE*, ABB2527DE*	Bottom Freezer	25.06	29.65	505	17.0
Maytag	MBF2556HE*, MBF2558HE*, PBF2555HE*, MB*2562HE*	Bottom Freezer	25.06	29.65	505	17.0
Amana	ABD2533DE*	Bottom Freezer	25.04	29.63	505	17.0
Kenmore	6328*30*, 6329*30*, 7328*30*, 7329*30*	Top Freezer	21.59	25.62	437	17.1
Kenmore	7325#, 7326#, 7525#	Top Freezer	22.14	25.92	445	17.2
LG Electronics	LRT*2232#**	Top Freezer	22.14	25.92	445	17.2
Amana	AFD2535DE*	Bottom Freezer	24.81	29.4	505	17.2
Kenmore	7550*, 7551*, 7552*, 7553*	Bottom Freezer	24.81	29.4	505	17.2
Maytag	MFD2560HE*	Bottom Freezer	24.81	29.4	505	17.2
Amana	AFB2534DE*	Bottom Freezer	24.79	29.38	505	17.2
Kenmore	7650*	Bottom Freezer	24.79	29.38	505	17.2
Kenmore	7651*	Bottom Freezer	24.79	29.38	505	17.2
Kenmore	7652*	Bottom Freezer	24.79	29.38	505	17.2
Kenmore	7653*	Bottom Freezer	24.79	29.38	505	17.2
Maytag	MFD2561HE*	Bottom Freezer	24.79	29.38	505	17.2
Maytag	MFF2557HE*	Bottom Freezer	24.79	29.38	505	17.2
Maytag	MFF2559HE*	Bottom Freezer	24.79	29.38	505	17.2

GODO STOYKE/CARBON BUSTERS

7.12: *The most efficient Energy Star fridges over 20.1 cu. ft. (sorted by consumption per cu. ft.). Based on EPA 2006 data and Internet research for retail pricing, efficiency based on adjusted volumes.*

nameplate on the back of the fridge, or from your old fridge manual (if you still have it). Then check Figure 7.14 to figure out how long it would take you to recover the cost of replacing your (working) older fridge with one of the less expensive, most efficient Energy Star fridges. There are two table sections in Figure 7.14: one for the northern US and Canada, and a second for the southern US.

Carbon Buster: 37.0%; $11,179.90

Brand	Model	Configuration	Volume	Adjusted Volume	kWh/ year	kWh/ ad cf
Sun Frost	R-19	Refrigerator Only - Single Door	16.14	16.14	204	12.6
Sun Frost	RF-12 p	To Freezer	10.12	11.41	171	15.0
Sun Frost	RF-16	Top Freezer	14.31	16.77	254	15.1
Monogram	ZIS480NR, ZISS480NRS	Side-by-Side	30.61	38.70	592	15.3
Kitchen Aid	KSSC48FM*0*, KSSO48FM*0*, KSSS48FM*0*	Side-by-Side	29.92	36.93	585	15.8
Kenmore	7420*40*, 7421*40*	Top Freezer	21.64	25.72	417	16.2
Profile	PTS25LBS, PTS25LHS, PTS25SHS, PTS25LHR*, PTS25SHR*	Top Freezer	24.60	29.10	475	16.3
Kenmore	7428*40*,7429*40*	Top Freezer	21.64	25.73	422	16.4
Kenmore	7554#, 7555#	Bottom Freezer	25.00	29.60	499	16.9
LG Electronics	GR-B258****, GR-F258****, LRFC2575#**, LRFD2585#**	Bottom Freezer	25.00	29.60	499	16.9
Amana	ABB2524DE*, ABB2527DE*	Bottom Freezer	25.06	29.65	505	17.0
Maytag	MBF2556HE*	Bottom Freezer	25.06	29.65	505	17.0

7.13: *The most efficient Energy Star fridges based on kWh consumption per cu. ft. (sorted by consumption per cu. ft.). Based on EPA 2006 data and Internet research for retail pricing, efficiency based on adjusted volumes.*

Quick Fact

"**coolth** (koolth) *n. Informal.* coolness, the state of being cool." Cool, eh?

NORTHERN PAYBACKS

You can see from the table that if the fridge is not too expensive, e.g.: the Summit 10-cubic-foot Energy Star fridge, you will recover the cost of replacing an ancient 1972 fridge in two years (annual savings around $150), a 1990 fridge in four years ($75 annual savings), and a 1993 fridge in seven years.

On the other hand, if you replace your old fridge with a larger model, e.g.: the 22 cu. ft. Kenmore, your paybacks are between 6 and 31 years. Finally, if you replace the old fridge with the GE Monogram (purchase cost $5,285 and up), you have one of the most energy-efficient fridges money can buy, rivaling the Sun Frost in efficiency, on a kWh per cu. ft. basis. However, due to its 30 cu. ft. size, the total consumption is still relatively high, and the purchase cost is so high that the energy savings pale by comparison.

SOUTHERN PAYBACKS

If you have significant air-conditioning bills, your payback from replacing your old fridge is faster.

The reason for this is that if you use air conditioning (A/C), you can save 30 percent in A/C bills in addition to your reduced fridge power costs, since the more efficient fridge will contribute less heat to your house. (On the down side,

Carbon Miser: 24.9%; $9,386.00

Brand Model	Summit CM-115ET9AHT*M*0*, ET9FHT*M*0*	Whirlpool GR9FHM*P*0*, GR9FHK*P*0*, GR9SHK*M*0*	Kenmore 7325#, 7326#, 7525#	Sun Frost RF-12	GE Monogram® ZIS480NR	
Type	Top Freezer	Top Freezer	Top Freezer	Top Freezer	Side-by-side	
Price	$289	$588	$750	$1,820	$5,285	
Cubic feet[1]	10	19	22	10	31	
kWh/year[2]	285	417	445	171	592	

	Average annual historic fridge energy	Simple payback[2] (Years)					Energy Return on Investment[2] (EROI, in years)
Year	consumption (kWh)						
Northern US/Canada							
1972	1726	2	5	6	12	47	1
1987	974	4	11	14	23	140	2 to 4
1990	965	4	11	15	23	143	2 to 4
1993	691	7	22	31	35	540	3 to 16
2006[4]	171 to 727						
Southern US[3]							
1972	1726	2	3	6	9	35	1
1987	974	3	8	14	17	103	2 to 3
1990	965	3	8	15	17	105	2 to 3
1993	691	5	16	31	26	397	2 to 12
2006	171 to 727						

7.14: *Paybacks for fridge replacement.*
The time it takes to recoup your investment (simple payback in years) when replacing a 1972 to 1993 fridge with select 2006 Energy Star model. Energy return on investment (EROI) shows time in years to make up the estimated energy lost in manufacture of fridge.
1 interior volume; 2 lower is better; 3 if you live in the southern half of North America, you will realize an additional 30% power savings from a more efficient fridge due to reduced air conditioning loads, accounting for faster payback times; 4 Energy star rated fridges only for 2006.

you will also get less heat from your energy-efficient fridge in winter. However, warmth is several-fold times cheaper to produce than coolth.)

For example, replacing your old fridge with a 19 cu. ft. Energy Star fridge will pay for itself in eight years, and replacing it with a 10 cu. ft. fridge will reduce that to three years (Figure 7.14).

Quick Fact

What is "Payback"?

Payback is the time it takes, in years, to pay for all the costs of implementing an energy investment out of the achieved energy savings. For example, if a refrigerator costs you $300, and saves you $100 per year in energy bills, the simple payback is three years. If the fridge costs $800, and saves you $100 per year in energy bills, the simple payback is eight years. ("Simple" payback means that the cost of borrowing money is not included in the calculation.)

Does it take more energy to make a new fridge than the new fridge will save?

In short, no. One study found that it takes 1,650 kWh to manufacture a fridge (2,100 pounds of CO_2).[65] Assuming that this figure includes the energy costs of the full supply chain, this means that a 1972 fridge consumed more power per year than it took to make it. Therefore, replacing the fridge with an efficient one has an energy return on investment (EROI) of a bit over one year in the north, under one year in the south.

And the most important energy-saving opportunity when buying a new fridge: don't move the old one into the basement or the garage. Otherwise, that old fridge may well be the highest power consumer in your house, and that lonely six-pack of beer in the basement may be the most expensive beverage you will ever drink!

Gas Refrigerators

Many off-grid homes use an alternative to electrical refrigerators: gas-powered fridges. These fridges are also popular in the RV world and are usually supplied with propane, though conversion kits to natural gas are available for about $75.

Access: warehouseappliance.com

Freezers

The most efficient freezers are all chest freezers (as opposed to uprights), and this is no coincidence; hot air rises, cold air sinks. Therefore every time you open the door of an upright freezer, a substantial amount of cold air flows out of the freezer. This is much less of a problem with a chest freezer, where little coolth escapes when the door is opened. The same applies to the door seal — there is much less leakage with the chest freezer.

Figure 7.15 lists the freezers with the lowest energy consumption (kWh) per year, while Figures 7.16 and 7.17 list the most efficient freezers based on kWh consumption per cubic foot of freezer space, for models up to 10 cubic feet and over 10 cubic feet, respectively.

Carbon Miser: 24.9%; $9,386.00

Brand	Model	Configuration	Volume	kWh/year	kWh/cu. ft.
Wood's	C09**E	Chest Freezer	9.0	251	27.9
Crosley	WCC10/E	Chest Freezer	10.0	282	28.2
Danby	DCF10**WE, DCF1014WE, DCF1024WE	Chest Freezer	10.0	282	28.2
Maytag	MFC10**AEW	Chest Freezer	10.0	282	28.2
Whirlpool	EH101*	Chest Freezer	10.0	282	28.2
Wood's	C10**E, C101**E	Chest Freezer	10.0	282	28.2
Avanti	VM799W	Upright Freezer	7.5	292	38.9
Crosley	WCC12/E	Chest Freezer	12.2	298	24.4
Wood's	C12**E	Chest Freezer	12.2	298	24.4
Absocold	AFD7501MW	Upright Freezer	7.5	341	45.5
Kenmore	24702400, 25702500	Upright Freezer	7.5	341	45.5
Sanyo	HF-7530*	Upright Freezer	7.5	341	45.5
Wood's	V10NAE*, V10W*E	Upright Freezer	10.4	353	33.9
Amana	AQC15**AEW	Chest Freezer	14.8	354	23.9
Danby	DCF15**WE	Chest Freezer	14.8	354	23.9
Danby	DCF1504WE	Chest Freezer	15.0	354	23.6
Maytag	MQC15**AEW	Chest Freezer	14.8	354	23.9
Whirlpool	EH151*	Chest Freezer	14.8	354	23.9
Crosley	WCC17/E	Chest Freezer	16.5	360	21.8
Wood's	C17**E	Chest Freezer	16.5	360	21.8

(right margin, rotated) GODO STOYKE/CARBON BUSTERS

7.15: *The most energy efficient freezers (kWh per year - lower is better; based on EPA 2006 data and Internet research for retail pricing, efficiency based on adjusted volumes).*

Brand	Model	Configuration	Volume	kWh/year	kWh/cu ft
0 to 10 cu ft					
Wood's	C09**E	Chest Freezer	9.0	251	27.9
Crosley	WCC10/E	Chest Freezer	10.0	282	28.2
Danby	DCF10**WE	Chest Freezer	10.0	282	28.2
Danby	DCF1014WE	Chest Freezer	10.0	282	28.2
Danby	DCF1024WE	Chest Freezer	10.0	282	28.2
Maytag	MFC10**AEW	Chest Freezer	10.0	282	28.2
Whirlpool	EH101*	Chest Freezer	10.0	282	28.2
Wood's	C10**E	Chest Freezer	10.0	282	28.2
Wood's	C101**E	Chest Freezer	10.0	282	28.2
Avanti	VM799W	Upright Freezer	7.5	292	38.9
Absocold	AFD7501MW	Upright Freezer	7.5	341	45.5
Kenmore	24702400	Upright Freezer	7.5	341	45.5
Kenmore	25702500	Upright Freezer	7.5	341	45.5
Sanyo	HF-7530*	Upright Freezer	7.5	341	45.5
Perlick	H1F	Upright Freezer	4.9	436	89.0

(right margin, rotated) GODO STOYKE/CARBON BUSTERS

7.16: *The most energy efficient freezers from 0 to 10 cu. ft. (kWh per cu. ft. per year - lower is better; based on EPA 2006 data and Internet research for retail pricing, efficiency based on adjusted volumes).*

Carbon Buster: 37.0%; $11,179.90

Brand	Model	Configuration	Volume	kWh/year	kWh/cu ft
Over 10 cu ft					
Wood's	C20**E	Chest Freezer	20.3	415	20.4
Maytag	MQC22**AEW	Chest Freezer	21.7	460	21.2
Whirlpool	EH221*	Chest Freezer	21.7	460	21.2
Wood's	C22**E	Chest Freezer	21.7	460	21.2
Wood's	C221**E	Chest Freezer	21.7	460	21.2
Crosley	WCC17/E	Chest Freezer	16.5	360	21.8
Wood's	C17**E	Chest Freezer	16.5	360	21.8
Danby	DCF1504WE	Chest Freezer	15.0	354	23.6
Amana	AQC15**AEW	Chest Freezer	14.8	354	23.9
Danby	DCF15**WE	Chest Freezer	14.8	354	23.9
Maytag	MQC15**AEW	Chest Freezer	14.8	354	23.9
Whirlpool	EH151*	Chest Freezer	14.8	354	23.9
Crosley	WCC12/E	Chest Freezer	12.2	298	24.4
Wood's	C12**E	Chest Freezer	12.2	298	24.4
Crosley	WCV17/E	Upright Freezer	16.9	430	25.4
United	UCF170/*E	Upright Freezer	16.9	430	25.4
Wood's	V17NAE	Upright Freezer	16.9	430	25.4
Wood's	V17W*E	Upright Freezer	16.9	430	25.4
Amana	AQU1525AEW	Upright Freezer	15.2	409	26.9
Crosley	WCV15/E	Upright Freezer	15.2	409	26.9

Godo Stoyke/Carbon Busters

7.17: *The most energy efficient freezers over 10 cu. ft. (kWh per cu. ft. per year - lower is better; based on EPA 2006 data and Internet research for retail pricing, efficiency based on adjusted volumes).*

Dishwashers

Very efficient dishwashers may use less energy than washing by hand.

Check table below for some efficient dishwashers. You can also take a look at the Energy Star website: energystar.gov.

Quick Fact

Energy Efficient Dishwashers. Source: ACEEE, E Source[66]

Brand	Energy use (kWh/year)	EF	Annual energy cost ($)
Equator	166	1.29	14
Asko	181	1.19	15
Viking	232	0.93	19

Note: EF = energy factor, higher is better with KWh, lower is better.

I hate housework. You make the beds, you do the dishes, and six months later you have to do it all over again."

— Joan Rivers

 Carbon Miser: 24.9%; $9,386.00

Quick Fact

Beware the hidden cost of electricity: each kWh of power requires 3 kWh of primary energy to produce it (US national average).

Cooking

Fuel Switching: Carbon Efficiency of Natural Gas Ranges

The most effective way to save on range cooking costs and carbon emissions is to switch your energy source: from electrical power to natural gas. The reason for this is outlined in "How to Benefit from Fuel Switching," page 22; electricity is the least efficient and most costly way of producing heat, since in the production of power two thirds of the energy is lost as waste heat.

Assuming you already have a natural gas hookup, the cost of installing an additional gas line may be around $150, depending on the accessibility of your kitchen. Savings from a natural gas range are around $42 per year. The range itself may cost $379 and up, comparable to an electric range (though you can spend much more on either, if you are so inclined). So, should your old range give out, a gas range will pay for itself in about three years. Make sure you buy a gas range with electric ignition. Gas ranges with pilot lights may use as much as 60 percent more energy over the year compared to gas ranges with electronic ignition. (The energy consumed by the electronic ignition itself is minimal.) All plug-in gas ranges sold in the US are required by law to have electronic ignition.

Keep in mind that the combustion products of natural gas include carbon monoxide and nitrous oxides. When the flame is burning correctly (blue color) the emissions are very small. However, emissions jump when the flame is yellow due to incomplete combustion. In that case, it is time to clean the burner or have your gas company check it out.

Sealed combustion units venting directly to the outside have also been developed to eliminate combustion products in the home.

Carbon Buster/Miser Recommendation: Replace electric stove with natural gas stove (new, if necessary).

5-year savings: $209, 4,100 lbs. CO_2, 17,500 cu. ft. (18.4 gigajoules) of natural gas replace 4,900 kWh of electrical power.

Life-time savings: $1,045, 20,400 lbs. CO_2, assuming life of 25 years.

Incremental cost: $150 for installation of gas line. **New cost:** $529; $379 for stove plus $150.

Payback incremental: 3.6 years. **IRR:** 28%. **CROI:** 136 lbs./$.
Payback new: 12.7 years. **IRR:** 6.1%. **CROI:** 38.5 lbs./$.

Carbon Buster: 38.4%; $11,389.00

GODO STOYKE/CARBON BUSTERS

Appliance	1° Energy (kWh)	CO_2 (lbs)	Cost (¢)
Gas frying pan	1.1	0.4	5
Electric microwave oven	1.2	0.5	4
Electric crockpot	2.1	0.9	7
Gas oven	2.4	0.9	12
Electric frying pan	2.7	1.1	9
Electric toaster oven	3.0	1.3	10
Electric convection oven	4.2	1.8	14
Electric oven	6.0	2.5	20

7.18: *Primary energy used to cook a meal, and the resulting carbon dioxide emissions and costs. Source: Northeast Utilities, ACEEE, Carbon Busters.*

GODO STOYKE/CARBON BUSTERS

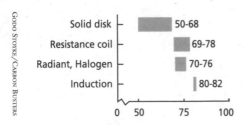

7.19: *Approximate efficiency of electric cooktop elements.*

What is the Most Efficient Cooking Appliance?

Figure 7.18 shows the efficiency of a variety of cooking appliances and cookware:

- Stoves are less efficient than cook-tops, because you are heating up a lot of mass before heating the food.
- Electric convection ovens are more efficient than regular ovens.
- Natural gas uses the least primary energy.

Types of Electric Stoves

There is variation in the efficiency of different electric cooktop models, but reliable data are scarce, and there is great variation from model to model, i.e.: among resistance coils (the most common), solid disks, radiant and halogen elements, and induction stoves. Overall, solid disks tend to be less efficient, induction elements slightly more efficient (Figure 7.19).

User Behavior

A study by the US National Bureau of Standards found that the energy required by different cooks to prepare the same meal using the same appliance varies by 50 percent.

Keep in mind that each time an oven is opened during baking, it causes a loss of 20 percent of the oven's energy content.

Energy-Efficient Cookware

Believe it or not, even your cookware can affect the energy consumption of your home. For example, a pot with a warped bottom can use 50 percent *more* energy than a flat-bottomed pot on an electric solid-disk stove-top. On the other hand, an insulated pan, or a pressure cooker, can drop the consumption by 58 and 68 percent, respectively, compared to a flat-bottomed pot (Fig. 7.20).

The Rocky Mountain Institute demos two particular kinds of efficient cookware at its headquarters building:[67]

 Carbon Miser: 26.3%; $9,595.00

One is the English Simplex, the Cadillac of teakettles. A specialized version for gas ranges increases heat transfer to the water by the use of a heat-retaining coil at its base that entangles hot air, thereby increasing the amount of heat transferred from the gas flame to the kettle. A copper and a chrome version are available, the latter not requiring any maintenance to keep it shiny. The kettle can save 25 percent of your hot water heating energy. A two quart kettle costs $90, or about 25 to 35 percent more than other kettles.[68]

Access: Simplex Tea Kettle: (25 percent energy savings) amazon.com. Simplex Kettle Co., 275 West Street #320, Annapolis, MD 21401

The other technology is the Swiss Rikon pot (Fig. 7.21). This stainless steel Durotherm thermal cookware has a double wall and double lid system that cuts your energy use by up to 60 percent. The thermally insulated pot keeps cooking the food even after it is removed from the stove and placed on the table, thereby greatly reducing cook time. The cookware is expensive, but costs no more than other high-quality cookware.

The pot provides a number of other benefits:

- The food is kept hot on the table for up to two hours.

- Waterless cooking: many foods can be cooked in their own juices, providing tastier, more vitamin-rich meals.

- Once removed from the stove, the self-cooking action prevents food from burning for worry-free cooking completion.

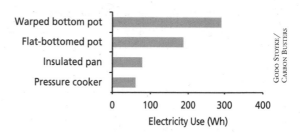

7.20: *Energy required to boil 0.4 gallons of water on an electric range dependent on choice of cookware (shorter is better; excluding primary energy consumption).* From E SOURCE.

7.21: *Efficient cookware, such as this double-walled Kuhn Rikon Durotherm cook and server pan, can cut your cooking energy costs by 25 to 60 percent.*

Quick Tip

Using energy-efficient cookware can cut your cooking energy costs by 25 to 60 percent.

Carbon Buster: 38.4%; $11,389.00

- The aluminum core, surrounded by stainless steel, distributes heat evenly over the bottom surface of the pot, greatly reducing the likelihood of burning during the heat-up phase.[69]

The Kuhn Rikon 2-Liter (2-quart) Durotherm Cook & Serve Pan is $ 169.

Access: Rikon double-walled cook pots: Available at local kitchen specialty stores, or at kuhnrikon.com or factorydirect2you.com.

Carbon Buster Recommendation: Purchase efficient cookware; double-walled cook pot and heat-entangling gas kettle (incremental).

5-year savings: $137, 1,060 lbs. CO2, 8,700 cu. ft. natural gas.

Life-time savings: $738, 5,700 lbs. CO2, assuming life of 27 years.

Incremental cost: $30 (assuming that you are in the market for high quality cookware). **New cost:** $259; $169 for cook pot, $90 for kettle.

Payback incremental: 1.4 years. **IRR:** 73 %. **CROI:** 152 lbs./$. **Payback new:** 9.5 years. **IRR:** 9.7 percent. **CROI:** 22 lbs./$.

Note: kettle savings only apply to gas stove. Double-walled cook pot dollar savings are about twice as high on an electric stove.

Two other technologies that reduce energy costs and are making a comeback are crock-pots and pressure cookers (Figure 7.20). Even simply using a flat-bottomed pot, as opposed to a warped bottom, can cut energy requirements by 35 percent (though the shape of the pot bottom has no effect on gas stove savings).

Solar Ovens

Solar ovens are revolutionizing cooking in many rural villages in developing countries around the world, replacing scarce firewood and reducing greenhouse gas emissions. Solar ovens use concentrated sunshine to cook food. While they are not widely used in Canada and the US, green architect Jorg Ostrowski uses a solar oven to prepare 75 percent of his cooked meals in wintery Calgary, Alberta (Fig. 5.4, p. 28).

If you have regular access to sunshine, you may want to try a solar oven for a novel cooking experience. While fancy solar ovens may cost as much as $249 realgoods.com, simple versions are available for as little as $20 solarcookers.org, or can be home-built solarcooking.org.

Clothes Washers

There are two basic types of washing machines: top- and front-loaders. Front-loading washing machines (also known as H-axis washers, due to their horizontal axis) are hugely more energy efficient.

Front-loaders are about three times more efficient than conventional, top-loading washers, both with respect to water and energy use. Front-loading washers clean better, and are far gentler on your clothes, resulting in hundreds

 Carbon Miser: 26.3%; $9,595.00

of dollars of savings on new apparel, in addition to the energy savings.

Also, good front-loaders achieve higher spin cycle speeds, resulting in drier clothes from the washing and therefore lower bills on drying, as well (or faster dry times, if you hang clothes to dry).

Many US states, cities, or utilities offer rebates for the purchase of highly efficient washers or dryers, so be sure to check in your area.

Savings: An average family has about 380 loads of laundry per year, using about 39 gallons of water per full load with a conventional washer. This accounts for nearly 15,000 gallons, or 15 percent of US average household consumption. A front-loading washer can save you over 60 percent of this water.

Additionally, you will save about $580 worth of detergent, and several hundred dollars worth of clothes over the life of the washer, which were not included in this payback calculation.[70]

This front-loader is more efficient, its spin cycle gets more water out of your clothes and it even hangs them up to dry!

Quick Tip

Front-loading washing machines use one-third the water and energy of top loaders, spin-dry your clothes better, can save you hundreds of dollars in clothing replacement costs.

  Carbon Buster/Miser Recommendation: Replace top-loading washer with front-loading model when shopping for a new machine.

5-year savings: $334, 3,300 lbs. CO2, 540 kWh of power, 13,000 cu. ft of natural gas and 37,000 gallons of water.

Life-time savings: $734, 7,200 lbs. CO2, assuming life of 11 years.

Incremental cost: $53. **New cost:** $614.

Payback incremental: 0.8 years. **IRR:** 127 %. **CROI:** 138 lbs./$.

Payback new: 9.2 years. **IRR:** 3.1 %. **CROI:** 11.8 lbs./$.

Note: If your hot water tank is heated electrically instead of with natural gas, your savings will be roughly *double* those indicated above (see p. 115).

If you don't want to spring for a new washer, you can achieve significant savings by reducing the temperature of your wash. Always turn the rinse cycle to "cold," since your clothes will not get cleaner by being rinsed with warm or hot water, and cold water is not as hard on your clothes.

Also try using warm or cold water for the wash cycle instead of hot water. Hot water tends to shrink your clothes and fades and wears your clothes out more quickly.[71]

Carbon Buster: 39.9%; $11,859.50

Carbon Buster/Miser Recommendation: Use warm/cold instead of hot/hot water setting on clothes washer. Note: calculations are for gas-heated water. Double savings for electrically heated water.

5-year savings: $74, 600 lbs. CO_2, 4,800 cu. ft. natural gas.

Life-time savings: $371, 2,900 lbs. CO_2, assuming application for 25 years.

Incremental cost: $0. New cost: $0.

Dryers

Electric dryers vary little in energy efficiency. However, natural gas dryers offer big savings over electric dryers, both in terms of money and environmental benefits. This is due to the fact that dryer energy consumption is mostly due to the need for heating energy (see "How to Benefit from Fuel Switching," p. 22)

If you are also installing a front-loading washer, subtract $51 from your five-year savings, as we have already credited 10 percent of dryer savings to the front-loading washer.

Carbon Buster/Miser Recommendation: Replace electric dryer with gas dryer (incremental).

5-year savings: $231, 4,500 lbs. CO_2, 5,400 kWh of power (minus additional consumption of 19,500 cu. ft. of natural gas).

Lifetime savings: $509, 9,900 lbs. CO_2, assuming life of 11 years.

Incremental cost: $200; $50 price difference for gas dryer, $150 for installation of natural gas connection. New cost: $600; $450 for dryer, $150 for installation of natural gas outlet.

Payback incremental: 4.3 years. IRR: 20%. CROI: 50 lbs./$.

Payback new: It is not worth replacing an electric dryer with a gas dryer for the energy savings alone (negative internal rate of return).

Computers and Home Office Equipment
Life-Cycle Carbon Costs of Computers

A fairly recent study of computer life-cycle costs determined that the manufacturing energy cost of the computers studied was about 2,600 kWh. This means that 3,400 pounds of carbon dioxide are released before you even switch on your computer for the first time, representing 40 percent of its life-cycle carbon emissions. The computer will use another 3,700 kWh of power during use, for a total life-cycle emission of about 8,000 pounds.

That's a lot of carbon for a device that may only weigh 20 pounds or less itself![72]

Choice of Computer: Laptop vs. Desktop

Laptop computers are the efficiency champs, hands-down (Figure 7.22). Though desktop computers are getting more efficient, the power consumption of laptops is lower by a factor of 3 to 6.

 Carbon Miser: 29.1%; $10,211.10

On the downside, laptops are more expensive, slower and less upgradeable.

However, laptops have gotten a lot cheaper in the last few years, and their specs out of the box fulfill all of the needs of 90 percent of all users (notable exceptions being designers, scientists and hard-core gamers).

2005 actually marked the year in which total laptop sales eclipsed sales in desktop machines — a clear sign that buyers appreciate laptop benefits such as smaller footprints and portability.

In an office, a typical laptop can save $65 in electricity per year, compared to a desktop machine. In our office at Carbon Busters, 80 percent of all computers are laptops, one of the reasons why our office consumption per square foot is 90 percent below the national average — in fact, our building now uses half the power it consumed when it was unoccupied before we moved in.

At home, your savings depend on your current usage pattern; if you normally leave the computer running full-time, or have a home office, your savings are just as high as at work. If you already turn your machine off, your savings potential from a more efficient computer is accordingly lower.

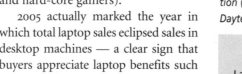

GODO STOYKE/ CARBON BUSTERS

7.22: *Typical laptop vs. desktop computer consumption (shorter is better; Extreme mini-tower with Daytek 15" LCD display vs. 14" Apple iBook).*

Quick Tip
Laptops beat desktops by a ratio of up to 6:1 in energy use.

Energy-Efficient Desktops

However, even among desktop machines there are significant consumption differences. While direct comparisons are not quite valid due to differences in processors and configurations, the variability in efficiency is striking (Figure 7.23). For example, Apple's Mac Mini (a desktop engineered like a laptop) consumes 13 watts during normal operations. By contrast, Dell's Precision 5300 (intended as a server) consumes slightly more (13.2 watts) *even while it is turned off!*

If we compare the Mac Mini with an Extreme PC entry-level system, which has similar performance, the difference is still remarkable: the Extreme PC uses almost three times as much energy *in sleep mode* as the Mac Mini uses while operating at full capacity (Figure 7.24).

Using the Computer's Sleep Mode

Setting your computer to sleep mode can reduce its power consumption by a factor of 5.

Typically, during sleep mode your screen will go black, the processor will go into reduced mode, and your hard-drive may spin down. Unlike shutdown,

Make	Model Features	Normal	Stand-by	Off
Apple	Mac mini 1.25GHz, Model: M9686LL/B	12.8	0.7	0.0
Gateway	E - 3400 PIII 1.0 GHz	28.8	20.4	2.4
Compaq	EVO, model: D5pD/P1.7/20j/8/2/256c/6 US	52.0	7.2	0.0
Apple	iMac PowerPC G5, 1.9GHz, Model: MA063LL/A with 17 inch LCD display	68.0	8.9	1.0
Dell	Dimensions 2400	72.0	3.6	1.2
Dell	Dimension 8250	90.0	1.6	0.0
Crystal System	(home build)	96.0	54.0	7.2
Apple	Power Mac G5, with 30 inch Cinema HD display	111.0	15.0	0.0
Dell	Precision 5300 intended as a server	144.0	13.2	13.2
Computer access: Westworld computers, Generation Electronics, Three Hat				

7.23: *Energy efficiency of computer models.*

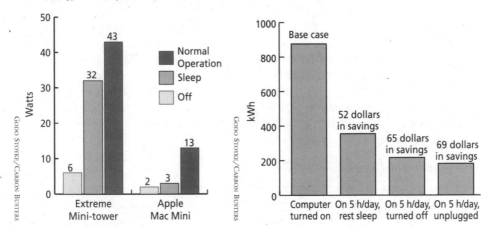

7.24: *Energy efficiency of two consumer desktop computers, excluding monitor, in watts of electricity; lower is better.*

7.25: *The effect of sleep and off modes on computer power consumption (100-watt desktop system and LCD monitor). Shorter is better.*

wake-up from sleep mode is instantaneous or takes only a very few seconds at most, depending on your model and operating system.

Figure 7.25 demonstrates the effect of operating your computer more efficiently, without paying a penny for a better computer model. The illustration is based on a typical desktop computer consuming 60 watts during operation, and 40 watts for a 17-inch LCD display. This computer will cost $87 to run for a year, leading to the release of 1,100 pounds of CO_2 during that time period. In fact, four of these computers would emit as much CO_2 as a Prius Hybrid Car would give off in the same amount of time!

Carbon Miser: 29.1%; $10,211.10

If your computer is programmed to enter sleep mode when not in use (assuming five hours of active use per day), your consumption drops by 521 kWh, with savings of $52 per year. Turning off the computer when not in use saves an additional $13, and unplugging the computer (or simply turning it off at the power bar), adds another $3.50.

If you add up these savings, you will find that they are enough to pay for a brand-new entry-level computer system every five years, with $40 left over for ice cream.[76] (Keep in mind, though, that manufacturing a new computer uses up 2,600 kWh of

Quick Tip

The energy savings from using sleep mode, and turning your computer off when not in use, can pay for your next computer system, in full!

Government Ratings Systems for Efficient Computers

The US Environmental Protection Agency (EPA) and the Department of Energy created a rating system for energy efficiency called "Energy Star" (energystar.gov). This system rates numerous products, from washing machines and cordless phones to houses, and assigns an "Energy Star" rating to products that exceed normal standards by specified amounts. The program has been a huge success, saving Americans $10 billion in energy costs in 2004 alone.[73]

The Canadian Government uses a similar program called EnerGuide to provide consumers with a rating system for appliances, windows, and many other products (oee.nrcan.gc.ca/energuide).

However, the Energy Star rating system for computers is inadequate, and the EnerGuide system does not rate computers at all.

Energy Star allows sleep mode consumption of 15 watts or more after 30 minutes of inactivity, and considers neither the power consumption during normal operations, nor the power consumption while the computer is turned off.

This means that if you work on your computer all day, with 30 minutes for lunch, and turn off your computer diligently at the end of each day, you may derive no benefit from an Energy Star rated computer whatsoever, since energy star only rates sleep mode performance.

This is not to say that the Energy Star rating is not useful, as there are still many computers that manage to consume more than 15 watts in sleep mode.

Yet, vast efficiency differences exist between computer models for normal operation as well (for an example, see Fig. 7.24).

Right now, a buyer's only options for determining the power consumption of prospective computers are to go to each manufacturer's web site (where information is usually hard to find, sometimes missing, or inaccurate), or to bring a power meter to the showroom. [74, 75]

Carbon Buster: 41.6%; $12,141.80

Quick Tip

Screen savers *do not* save energy! If you love your screen saver and are loathe to part with it, set it to kick in five minutes before sleep mode sets in. That way, you'll enjoy the best of both worlds.

energy. Upgrading the old one, if feasible, is better.)

Setting sleep mode for Windows XP, 2000, Me, 98, 95

(a) *Option 1*: Go to the "Start" button and select the "Control Panel." Select "Classic" view, then double click on "Power Options." Proceed to step (b) Option 2: Right-click anywhere on the desktop. Select "Properties" from the menu that appears. The "Display Properties" control panel will appear. Click on the "Screen Saver" tab, then click the "Power" button found in the "Energy-Saving Features" of the Monitor section.

(b) In the "Power Options Properties" window, select the "Power Schemes" tab, go to "Turn Off Monitor" and select appropriate intervals (e.g.: 10 minutes) for putting your monitor to sleep.

(c) In the "Power Options Properties" window, choose an appropriate interval to power down your hard drive under "Turn Off Hard Disk."

Another option is "System Standby." You determine an appropriate interval after which the computer goes into standby mode.

Setting sleep mode for Macintosh, OS X:
You can change settings here:
Apple Menu: System Preferences: Energy Saver: Sleep
Switch to "Options" to change settings for processor performance. "Automatic" is best, unless you play certain games that will display jerky behavior under this setting. The automatic setting will reduce the processor speed (and therefore its consumption) whenever no activity is detected. In fact, the system is so responsive that it will save energy even in the inactive time between keystrokes.

Note that you can set different settings while connected to the power adapter vs. running on battery (laptops only). Generally, you will want to be even more conservative while running on battery, to increase battery time remaining.

Suggested settings (while connected to power adapter): computer sleep 30 minutes, display sleep 10 minutes, put hard drive to sleep when possible.[77]

Setting sleep mode for Linux
Sleep mode for GNOME
Grab a command line and run: gnome-control-center as the user who is normally logged into X, rather than as root.

 Carbon Miser: 29.1%; $10,211.10

Choose "Advanced" and then "Screensaver," then click the "Advanced" tab where you will find the "Display Power Management" section. Adjust "Standby After," "Suspend After" and "Off After" settings to taste. Be sure to toggle on "Power Management Enabled."

Sleep mode for KDE

Simply run kcontrol and choose "Power Control/Display Power Control." From here, you can configure "Standby," "Suspend" and "Power Off" settings for your monitor.

There are numerous other tweaks and settings possible providing control over CPU frequency, hard-drive sync frequency and power-down timeouts. You can reduce how often the disk is used and hence increase the amount of time it can be off, by disabling atd and crond and optimizing (or disabling) the logging done by sysklogd. Amazing parsimony is possible, but at the cost of learning the ins and outs of Linux. Enjoy![78]

Access: linux.org/docs/ldp/howto/Ecology_HOWTO/ecology_howto_power_management.html

Carbon Buster/Miser Recommendation: Change your computer settings to automatic sleep-mode, turn off the computer altogether when it is not in use.

5-year savings: $343, 4,400 lbs. CO2, 3,470 kWh of power.
Life-time savings: $1,716, 22,000 lbs. CO2, applied over 25 years.
New cost: $5 for a power bar.
Payback new: 0.1 years (1 month). **IRR:** 1,373 %. **CROI:** 4,406 lbs./$.

Will I damage my computer if I turn it on and off?

Hard-drives built before 1984 were fragile things. So fragile, indeed, that some computer specialists recommended leaving the computer running day and night to reduce the stress of turning the machine on and off. Fast-forward to 2006: with improved hard drives and electronics, computers are now designed to withstand frequent on-and-off cycling, especially laptops. Hard drive mean-time-between-failure (MTBF) is more likely determined by head-disk mechanical interaction than electrical surges and thermal cycling.

In fact, computers that are turned off or in sleep mode are now expected to last longer, since they stay cooler.

Rule of thumb: Any computer built after 1984 can be safely turned off (or put in sleep mode) to maximize energy savings.

PS: Always use the shut-down command when turning off your computer.

Sources: RMI, Lawrence Berkeley National Laboratory, Home Energy Magazine, US Department of Energy[79]

Palmtops

Palmtops, notably Palms, Pocket PCs and Blackberries (and even smartphones), are gaining increasingly in capabilities. Attached to portable (and foldable) keyboards, and with the ability to read PDFs and read and write to word processing files, data bases, and spreadsheet documents, palmtops are becoming capable replacements for laptops while on the road. Few people have yet forsaken their laptops altogether for a palmtop, but the writing is on the wall. Running for a day (or days) on a single battery charge, palmtops can provide an efficiency boost by a factor of 100 or more compared to regular computers. Once LCD visors come down in price, these devices may become an important component of energy-efficient wearable computing.

Access: palm.com, palmsource.com, pocketpc.com, blackberry.com

Monitors

LCD (liquid crystal display) monitors are rapidly replacing CRT (cathode ray tube) screens as the display of choice for computers (and televisions, for that matter). LCD screens are a clear example that price is not always the driving factor in consumer demand. LCD screens actually have inferior color rendering capabilities, and color graphics professionals still mostly use CRT monitors for print applications. However, on the plus side, LCDs take far less space, are more portable, do not flicker, emit no electromagnetic radiation, and can be more energy efficient by a factor of 2 to 4.

The US government's Energy Star web site unfortunately does not list monitor energy consumption during normal operation.

Figure 7.26 shows the energy consumption in milliwatts (mW) per square inch for a number of monitors. The highest consumption (not surprisingly) is

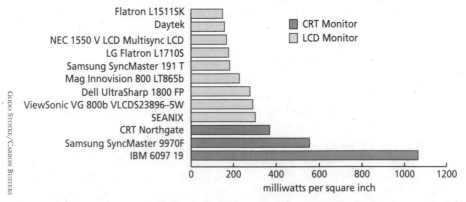

7.26: *Energy efficiency of LCD and CRT monitors during normal operation (milliwatts per square inch; shorter is better).*

Carbon Miser: 30.6%; $10,554.30

by a CRT, here the IBM 6091-19, and the lowest by an LCD, the Flatron L1511SK. LCD panels vary by as much as a factor of 2 in energy efficiency.

Bring your own power meter when buying a monitor, or ask the sales people for energy data, to get them thinking about energy efficiency.

Printers
INK-JET PRINTERS
Ink jets are very energy-efficient printers for low-volume printing, generally consuming little power during operation, and even less during stand-by.

As with most electronic devices, ink jets are best turned off with a power bar when not in use, as a typical home printer can use nearly ten times as much power per year while off, compared to actual use.

Cartridge Refilling
The most expensive fluid on the planet is the ink for an ink-jet printer; only expensive French perfume and a few select bottles of wine rival it in cost, on an ounce-per-ounce basis. Printer companies like HP, Canon, Epson or Lexmark tend to offer the printers themselves at very low cost, but make their real money on the cartridges. The average cost of the ink is around $1,700 per quart, by one calculation, explaining the global $21 billion ink market.

It also partly explains why printer manufacturers are constantly releasing new cartridge types: this makes it harder for refill businesses to keep up with the new formats.[80]

It may also explain why many printer manufacturers eagerly offer free recycling services for their cartridges: every recycled cartridge is one less cartridge that can be refilled.

Some manufacturers (e.g. Lexmark) have even gone to the length of providing lockout chips that will prevent the use of third-party cartridges, or providing cartridge rebates up front if the buyer agrees to not refill the cartridge.[81]

If you encounter such practices from a manufacturer, my recommendation is to vote with your feet, and walk.

While recycling is good, reuse is better. Rather than throwing away all the embodied energy of these little high-tech devices, a reused cartridge has a much longer serviceable life, allowing five or perhaps six reuses.

So wasteful is the practice of throwing out single-use cartridges that the European Union has banned the sale of non-refillable printer cartridges in the European Market.[82]

A review in PCWorld examined the quality of ink refills. While some refill inks were as good as the originals, others were not, and some even clogged cartridge nozzles.

Based on simulated, accelerated aging, the review also found that all tested ink refills faded after the equivalent of five years, while original inks lasted up to 92 years.[83]

It begs the question why ink cartridge manufacturers don't offer refills themselves, which would provide quality assurance for buyers.

Manufacturers may state that using refilled cartridges voids the printer warranty. However, the savings from refilled cartridges will probably pay for the cost of the ink-jet printer several times over.

Also, note that many ink-jet manufacturers now ship with cartridges that have only 50 percent capacity. When this cartridge is empty, you may want to purchase a full-capacity cartridge that is worth refilling.

We used to refill ink-jet printers ourselves in the 1990s. It was messy, but saved two thirds of the cost. We never had any problems, and the documents we printed 11 years ago do not seem to have faded. Then we started using a refill service provider imageres.com, which eliminated the messiness and still saved 50 percent of the cost.

For a number of years now we have only used laser printers with toners, which we also get refilled, again for about 50 percent savings. On the rare occasions (once or twice in six years) that there was a problem with the refilled cartridge, the service provider promptly provided a free replacement, including drop-off.

Recently, we started using a wax printer, eliminating cartridges altogether (on the down side, the wax printer needs to be left on at all times, or wastes an expensive dab of toner with each on/off cycle, definitely not an ideal solution).

My recommendation is to find a refill service provider or brand that works for you, and to stick with it. Ask friends or associates if they can recommend a reputable firm in your area.

For prints that you want to have around for a long time, use the manufacturer's ink, or find a refill business that guarantees the archival qualities of its inks.

Laser Printers

Laser printers use much higher wattages when printing than ink-jets. However, they are more suitable for high-volume, high-speed printing and have much lower per-page costs for consumables.

Energy Star-rated laser printers will enter sleep-mode in five minutes or less and use 10 watts of power or less in that mode. (Criteria are incrementally less stringent for higher volume, and for color printers.)[84]

According to Energy Star, over its lifetime the qualified equipment in a single home office (including computer, monitor, printer and fax) can save enough power to light an entire home for more than four years.

Consider a duplexing printer, if you are in the market for a laser printer. Duplexing units cost more, but according to Energy Star they will save you up to $30 per year in paper costs, based on their ability to print on both sides of a page automatically.[85] (Also, you will reduce carbon emissions: two pounds of virgin paper lead to the release of 7 pounds of CO_2.)[86]

 Carbon Miser: 30.6%; $10,554.30

Fax Machines

Most fax machines used in home offices are now ink-jet printers (or multi-function ink jets), so conservation tips for sleep-mode, Energy Star ratings and cartridge refills apply to them just as they do for ink-jet printers (see "Cartridge Refilling," p. 91). Unlike printers, though, fax machines have to stay on non-stop to be effective.

One way to avoid the stand-by energy consumption is the use of a "Fax Saver," a device which turns off power to the fax until an incoming call is detected.

However, using our 6.15-watt sleep-mode HP 925xi fax machine as a basis, this will only save you about $5.35 per year (plus a respectable 68 pounds of CO_2).

Business opportunity: At the time of writing, no North American vendor for Fax Savers had been found.

Modems, Hubs and Routers

These devices typically do not use much power (4 to 15 watts), though they do add up if you have several, and leave them on continuously.

The most effective solution is to place your cable or DSL/ADSL modem, ethernet hub or router on the same power bar as your computer. If you frequently access the network from your laptop in the living room, bedroom or kitchen, and don't want to run all the way to the basement to turn on your wireless router, consider putting your hub on a timer for times when you are typically sleeping or out of the house. This will also reduce the time window for illicit users trying to break into your network.

Some Internet service providers (ours, for example) prefer that you leave your high-speed modem on day and night for possible modem software updates. However, our system is always switched off via timer at night, and we are not aware of any problems due to this. If you are concerned about it, check with your Internet service provider. Most updates occur at regularly scheduled intervals to minimize interruptions (e.g.: Tuesday at 4 am), and you could occasionally leave the modem on for those nights, for the occasional update.

Photocopiers

Photocopiers differ greatly in their power consumption during operation, stand-by, and in the "off" position. However, since most copiers in homes will not see huge volumes of printing, the most important components of their consumption are the stand-by and off settings.

Low-volume Energy Star copiers (copiers that print 20 copies per minute [cpm] or less) must go into off-mode after no more than 30 minutes and may consume no more than 5 watts in that state (though there is no requirement to be ready to copy within 30 seconds, as there is for copiers that deliver 21 to 44 cpm).

This Energy Star requirement will have a big impact on your printer's power consumption; I have tested non-Energy Star printers where the stand-by mode

Carbon Buster: 43.1%; $12,484.90

had virtually no effect on power consumption. The Energy Star qualification will ensure that your copier has an effective stand-by function.

Keep in mind, though, that, as with other power vampires (see "Power Vampires," p. 67), turning your printer off does not mean it is not using power. In fact, of the dozens of larger office copiers we have tested, only two had no power draw in the "off" mode.

(Our office's Canon PC108OF photocopier has the distinction of not only having low power consumption [16 watts normal, 2.5 watts stand-by, 0 watts off], but of also being ready within one second after being woken from stand-by mode.)

> **Quick Fact**
>
> Make double-sided copies, or redeploy one-sided scrap paper for internal use; it takes ten times as much energy to make a piece of paper as it takes to copy onto it.
>
> Source: Energy Star[87]

Copiers with sorting trays typically draw about 35 watts continuously when they are nominally "off", and one model tested drew as much as 360 watts. Based on eight hours of daily use, 200 days per year, these copiers will cost you $25 and $255 per year, respectively, during the 16 hours per day they are "off." Fortunately, copiers with automatic sorting trays are rare in home offices.

If you use a multi-function ink jet for copying, follow the same recommendations given for ink-jet printers for energy savings (see "Ink-Jet Printers," p. 91).

Entertainment Electronics
Television

Televisions are getting larger and more numerous in homes. A positive trend for energy consumption, though, is the rapid move towards energy-efficient LCD TVs, a delayed mirroring of trends in computer monitors (Figure 7.27).

Larger TVs tend to use less energy per square inch of viewing area — yet some of the larger TVs use up more energy when they are turned *off* than some smaller ones use when they are running.

TVs are the classic power vampires; most television sets currently in our homes draw power even when they are turned off. Most TVs use only 2 to 6 watts when they are turned off, but when you add this up, two large power plants have to run day and night just to supply the energy to North America's television sets while they are *turned off.*

Satellite Receivers and Cable Boxes

Satellite receivers present a continuous draw of power, even when they are turned off. In fact, our StarChoice receiver drops by only 1 watt (from 27 watts to 26 watts) when it is turned off.

Carbon Miser: 30.6%; $10,554.30

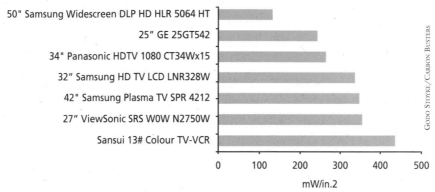

7.27: *Energy efficiency of television sets.*

This appears to be a universal problem with set-top boxes.

By unplugging the receiver at the same time as the television set, you will save $20 over the course of a year (based on three hours of TV per day).

One disadvantage of having your receiver unplugged is that it takes a while to load the program guide. If this is a concern for you, put your receiver on a timer for your typical viewing times, so that the programs are pre-loaded by the time you are ready to watch.

Quick Fact

Two massive power plants are running day and night to supply the power for our television sets — *while they are turned off!*

Christmas Lights

If you are looking for efficient Christmas lights, the choice is simple: LEDs (see "LEDs," p. 62). In this application, LED lights are more than 100 times more efficient than standard mini-bulb decorative light chains, since the LEDs provide a point-source application of light, as opposed to providing general illumination.[88]

 Carbon Buster/Miser Recommendation: Replace your conventional Christmas lights with LEDs.

5-year savings: $106, 1,400 lbs. CO_2, 1,100 kWh of power.

 Life-time savings: $1,056, 13,600 lbs. CO_2, assuming life of 50 years (nominal life of LED lights is 278 years, if left on 12 hours for 30 days every Christmas).

New cost: $60 for 2 strings of 75 LED lights each.

Payback new: 2.8 years. **IRR:** 35.2 %. **CROI:** 226 lbs./$.

Green Power

Green power is defined as energy derived from renewable energy sources whose production does not add net carbon dioxide to the atmosphere, though there

Carbon Buster: 43.6%; $12,590.50

are always carbon emissions during facility construction. Wind is one of the most common green power sources. Others include solar, biomass, biogas and low-impact hydro.

By buying green power, you help promote the construction of new sustainable power facilities. While a lot of green power is already cost-effective today, even without counting its environmental economic benefits, sometimes green power needs just a little extra push to edge out polluting fossil fuel sources.

By buying green power you can also make a very significant contribution to reduce carbon emissions. The critical point here is to check whether your green power provider is actively engaged in installing new green capacity to displace polluting power sources. Federal standards give you some assurance of its green rating, but it never hurts to check yourself by doing a bit of Internet research. Mind you, you do not stand to gain a cent from buying green power yourself. However, dollar for dollar, it is one of the most effective steps you can take to reduce your global carbon footprint.

The average incremental price for green power is 0.77 cents/kWh in the US, and CDN 2 cents/kWh in Canada (i.e., this is what you would pay on top of your regular power rates).[89]

Access: To find green power providers in your area go to:

US: epa.gov/greenpower/locator/index.htm

Canada: pollutionprobe.org/whatwedo/greenpower/consumerguide/c2_1.htm

Carbon Buster Recommendation: Replace conventional power with green power.

5-year savings: $0, 43,400 lbs. CO_2, 34,200 kWh of regular power replaced with green power.

New cost: $53 per year.

Payback: This measure currently has no personal financial payback. However, there are few ways to spend $53 that benefit the planet more. **IRR:** N/A. **CROI:** 4,123 lbs./$.

Renewable Energy Sources: Making your Own Green Power

You can produce your own pollution-free energy. The cheapest and easiest form of renewable energy is passive solar heating (see "Green Heating," p. 107), and green cooling (see "Green Coolth," p. 117), followed by geothermal (see "Geothermal," p. 113) and solar hot water (see "Solar Collectors," p. 115).

Producing green electricity is a bit more involved. Solar electricity is often the only power source that you can reasonably produce in the city — wind generators generally do not conform to urban bylaws, though there are exceptions (e.g.: see the Toronto Wind co-op, windshare.ca). Other forms of green power, such as micro-hydro, or producing energy from biogas, tend to be tied to specific rural locations.

 Carbon Miser: 31.1%; $10,659.90

Solar Electric Power: Photovoltaics (PV)

All you need for solar electric power production is a location to site the panels that gets good southern exposure. The solar access from 10 am to 2 pm especially has to be unobstructed, as most of the power is produced during this time (Figure 7.28).

Most solar panels tend to create power for 12-volt systems (they are typically rated as producing 16 to 17 volts, to be able to effectively charge 12-volt batteries). A few panels are capable of supplying to either 12-volt or 24-volt systems, as the higher-efficiency 24-volt systems are becoming more common.

Larger panels almost always cost less per watt than smaller panels (e.g.: it is cheaper to buy one 120-watt panel than three 40-watt panels; Figure 7.29).

You can run a number of lights and appliances directly off the 12- or 24-volt panels, in which case the only other things you would need are wiring, a mounting system, and perhaps a charge controller.

Sunshine is a form of energy, wind and sea currents are manifestations of this energy. Do we make use of them? Oh no! We burn forests and coal, like tenants burning down our front door for heating. We live like wild settlers and not as though these resources belong to us."

— Thomas Edison, 1916[90]

Keeping up with the *New* Joneses

photovoltaics

passive solar

hybrid car & bicycles

rain water catchment

large veggie garden

naturalized lawn

geothermal

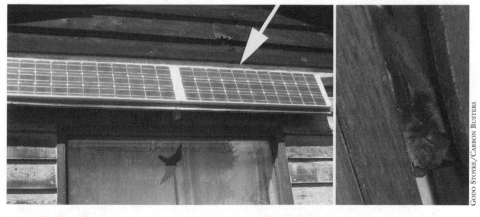

7.28: *Solar bat refugium: Sometimes this Little Brown Bat* (Myotis lucifugus) *disdains the bat box in favor of roosting under the two 48-watt photovoltaic panels.*

GODO STOYKE/CARBON BUSTERS

Carbon Buster: 58.8%; $12,590.50

Model	Price	Watts	Volts	$/W	warranty
Sharp 175W PV Module	879	175	12	5.02	25
Evergreen Cedar Series PV Module EC-115	585	115	12/24	5.09	25
Sharp 167W PV Module	870	167	12	5.21	25
Sharp 165W PV Module	875	165	12	5.30	25
Kyocera KC80 PV Module	439	80	12	5.49	25
Sharp 123W PV Module	699	123	12	5.68	25
Shell Powermax Module	940	165	12	5.70	N/A
Kyocera KC120 PV Module	685	120	12	5.71	25
Sharp 80W PV Module	495	80	12	6.19	25
Uni-Solar Triple Junction-US64	419	64	12	6.55	20
Sharp 70W PV Module	465	70	12	6.64	25
Kyocera 40W PV Module	275	40	12	6.88	25
Uni-Solar Triple Junction-US42	289	42	12	6.88	20
Kyocera 50W PV Module	345	50	12	6.90	25
Uni-Solar Triple Junction-US32	239	32	12	7.47	20

7.29: *Costs of photovoltaic solar panels (source: realgoods.com).*

GODO STOYKE/CARBON BUSTERS

Quick Fact

What types of solar panels are there?

Most people use the term "solar panel" a little loosely. Each solar panel actually fulfills one of two very distinct roles: it is used either to make electricity (photovoltaic panels, or PVs), or to make hot water (solar thermal water heating systems, solar thermal collectors, or simply solar collectors).

If your needs are modest, you can assemble a system for your weekend cabin for a few hundred dollars.

If you want to use the panels to run 120-volt appliances, you will also need an inverter which transforms the 12 volts into 120 volts.

Your next choice is whether you want to be grid-connected, or run independently off batteries; if you live in a rural area, the cost of bringing power to your house may be as much as an entire small PV system might cost. Plus, if you are not connected to the electrical grid, you will save fixed monthly utility fees, as well as charges for the wire service provider, which can easily be half of your monthly power bill. Keep in mind that with the battery-based, off-grid systems, there are losses of about 8 percent when converting battery power to AC power, and standby losses of about 1 percent per day of battery-stored power, thereby reducing the amount of usable energy available to you.

With a grid-connected system, you will pay all the base fees, regardless of how much power you send into the utility system. Also, if the utility power fails,

Carbon Miser: 31.1%; $10,659.90

so does your household power. On the plus side, you don't have to worry about times of inadequate sunlight (or wind), and you don't need to buy batteries.

Virtually all of the energy-efficiency measures described in this book offer better payback than a grid-connected solar PV system. However, there is something very satisfying about making your own power and, once all other options have been exhausted, it is a way to further reduce your emissions.

There has been some argument as to the net energy benefit of photovoltaic panels. The US Department of Energy is quite unequivocal in its position on PV: "Based on models and real data, the idea that PV cannot pay back its energy investment is simply a myth. Assuming 30-year system life, PV systems will provide a net gain of 26 to 29 years of pollution-free and greenhouse-gas-free electrical generation. This *includes* the energy to make the aluminum frame and the energy to purify and crystallize the silicon."[91]

For pricing of renewable power systems (wind and solar), see "Pricing a Renewable Power System," p. 101.

Wind

On a kWh by kWh basis, wind often offers better returns than photovoltaic panels. Of course, the payback depends on the amount of wind (and sunshine, in the case of PV panels) in your area.

Small differences in the speed of wind lead to major differences in power production. This is because the power produced increases with the cube of the wind speed. For example, when the speed of the wind increases from 5 meters per second (m/s) to 6 m/s, the power output almost doubles!

You can check a wind resource atlas at rredc.nrel.gov/wind to see if you live in a good wind area (windatlas.ca in Canada).

The higher the wind turbine tower, the more energy you get. It is generally not recommended to attach your wind turbine to your house. For one, your house is likely not high enough for optimal power output. For another, the vibrations transmitted from the wind turbine to your roof can be an unpleasant disturbance at higher wind speeds.

Generating electricity from wind in the city is not for the faint of heart: you will probably have a difficult time getting a permit.

For rural settings, wind complements solar nicely; solar tends to produce more power in the summer, wind more in the winter, making these two energy sources an excellent hybrid system for living off the grid.

Inverters and Batteries

Expect to pay about $1,400 for a high quality inverter, and at least $640 for a bank of eight reasonably good (deep-cycle, not automotive) 100 Amp-hour lead-acid batteries. (Note: Though the battery system described above is twice as

Carbon Buster: 58.8%; $12,590.50

large as the one we use in our home, it is still of a very modest size. For greater convenience, you can double or triple the number of batteries. However, you will of course have a respective increase in costs.)

You may also need a gasoline ($1,000 to $2,500) generator as a backup system, should there not be enough wind or sun at any time in the year, though some purists disdain the use of a fossil fuel backup generator as cheating (though the use of a biodiesel generator would be an environmentally friendly alternative).

Grid-Connected Systems: Net Metering

If you are going to be grid-connected, you need an intertie system that ensures your system goes off-line when the utility power is down (to ensure the safety of utility repair crews), a meter capable of going in reverse (to give you credit for your power contribution), and a cooperative utility company. The latter is vital; some utility companies will charge you so much money in annual fees for your connection that you cannot recover your costs, ever. The only option you have in this case is to be an "eco-guerrilla," and feed your excess power into the system without recompense, for the good of the environment.

On the other hand, some utilities positively encourage their customers to feed micro-power into the grid, and you will get as much money for your power as you would have paid the utility for it.

Fortunately, many state and provincial laws are slowly changing to recognize that regulations need to be simplified to allow hassle-free, grid-connected contributions from renewable micro-power, while some jurisdictions even pay handsomely for renewable power, recognizing its real economic and environmental value (for example, Ontario recently announced that it will start paying 42 cents per kWh for solar electricity to select groups, to encourage renewable generation.

Renewable Tax Credits

Tax credits are available for renewables for many areas of the US (14 states) and Canada. They range from $500 in Montana to $20,000 over four years in Idaho. The tax credits are available for solar hot water collectors, photovoltaics, wind, geothermal, biomass and microhydro.

Though there were no residential solar federal incentives in either Canada or the US in 2005,[92] this has changed for 2006 and 2007 (at least in the US).

The US Energy Policy Act of 2005 gives homeowners a tax credit of 30 percent for qualifying solar power (PV) or solar water heating costs, up to a maximum of $2,000 per system. These credits apply to equipment installed in 2006 or 2007.[93]

You can get a free guide to the tax credits from the Solar Energy Industries Association here: seia.org.

For further information on locally available incentives (for example the Chicago Solar Partnership), check out ecobusinesslinks.com.

 Carbon Miser: 31.1%; $10,659.90

In Canada, there is a Provincial Sales Tax Rebate on all solar equipment sold in Ontario. (Canadian Solar Industries Association, cansia.ca/government.htm).

Pricing a Renewable Power System

What are the costs for a modest wind or solar electric installation for your home? Below we examine a few different types of possible installations.

For tips on installation, or referrals to reputable local solar dealers and installers, you may want to contact your local branch of the Solar Energy Society, where available (US: ases.org, Australia and New Zealand: anzses.org, UK: thesolarline.co.uk).

For our calculations it was assumed that you have average hours of sunlight (i.e., you live neither in Las Vegas, nor Alaska — if you do, assume higher outputs from PV panels in Vegas, lower outputs in Alaska) and moderate wind speeds.

All of the systems described are modest. Some individuals pay $30,000 or more for large PV systems. However, we have assumed that you will take advantage of the other, more cost-effective efficiency options described in this book first, to lower your electrical consumption. This makes the most economic sense. If you *do* install a $30,000+ system, it is much harder to achieve paybacks (unless you have a preferred feed-in tariff system for small-scale renewables in your area).

CITY-BASED, GRID-CONNECTED SOLAR ELECTRICITY (PV) INSTALLATION

The grid-connected solar electricity option involves the least change in your household; a portion of your electrical consumption is simply supplied by PV panels, or fed into the grid in times of excess capacity (assuming your electrical provider is amenable to this net-metering method). The payback for this system is 35 years. Nominal power output: 1,050 watts.

OFF-GRID, RURAL SOLAR ELECTRICITY (PV) INSTALLATION

For the off-grid version, it is assumed that you have achieved some major efficiency gains in your house, such as those outlined in this book: it is far cheaper to buy efficient home-energy equipment than it is to supply inefficient equipment with more solar panels. A typical off-grid installation might include propane (or natural gas,

Item	Number	Cost/unit	Subtotal
Sharp 175W PV Module	6	$879	$5,274
Mounting brackets	6	$90	$540
Charge Controller	1	$250	$250
Intertie Inverter Sunny Boy 2500 w/ Display	1	$1,250	$1,250
Fuses, wiring	1	$150	$150
Labor	20	$50	$1,000
Subtotal			$8,464
US federal renewable tax credit			-$2,000
Total			$6,464

7.30: *Grid-connected PV system.*

Carbon Buster: 58.8%; $12,590.50

Item	Number	Cost/unit	Subtotal
Sharp 175W PV Module	6	$879	$5,274
Mounting brackets	6	$90	$540
Charge Controller	1	$250	$250
Deep-cycle batteries	8	$80	$640
DC to AC Inverter, 2.4 kW, Xantrex DR2412	1	$1,395	$1,395
Back-up generator	1	$1,150	$1,150
Fuses, wiring	1	$150	$150
Labor	12	$50	$600
Subtotal			9,999
US federal renewable tax credit			-$2,000
Total			7,999

7.31: Off-grid PV system.

Item	Number	Cost/unit	Subtotal
Whisper-200/1000watt	1	$2,000	$2,000
Tower	1	$1,000	$1,000
Charge Controller	1	$250	$250
Deep-cycle batteries	8	$80	$640
DC to AC Inverter, 2.4 kW, Xantrex DR2412	1	$1,395	$1,395
Back-up generator	1	$1,150	$1,150
Fuses, wiring	1	$150	$150
Labor	12	$50	$600
Total			$7,186

7.32: Off-grid wind system.

if available) to supply range, fridge and hot water, and a backup generator for periods of long cloud cover or winter weather spells.

The kWh output is rated for the central US: if you live further south, you will have higher power output. If you live further north, it will be less. The further north you live, the more a PV/wind hybrid system makes sense: solar electricity is abundant in summer, and wind energy tends to be more prolific in winter (see "Wind," and "Wind/Solar Hybrid," below).

With this system, you will need to supply some electricity from a generator, especially in winter. The payback for this system is 7.2 years, assuming a $6,000 cost if you had connected your house to the grid. Nominal power output: 1,050 watts.

Off-Grid, Rural Wind Power Installation

With this system, you also need to supply some electricity from a generator, in times of low wind. The payback for this system is 3 years, assuming a $6,000 cost if you had connected your house to the grid. Nominal power output: 1,000 watts.

OFF-GRID, RURAL WIND/SOLAR HYBRID POWER INSTALLATION

This scenario combines the best of both worlds: solar and wind electricity are often complementary, and should provide you with a fairly even supply of electricity. The payback for this system is 13.5 years, assuming a $6,000 cost if you had connected your house to the grid. Nominal power output: 2,050 watts.

 Carbon Miser: 31.1%; $10,659.90

Recommendation: Set up a rural, off-grid solar/wind hybrid power installation.

5-year savings: $2,640 (includes $180 savings per year of base fee connection charges), 22,749 lbs. CO_2, 18,300 kWh of conventionally produced power, minus 25 additional gallons of gasoline required for occasional backup power and battery conditioning. You must implement the other power saving recommendations outlined in this book, and possibly a few more, to be able to operate with the proposed renewable energy system.

Item	Number	Cost/unit	Subtotal
Sharp 175W PV Module	6	$879	$5,274
Mounting brackets	6	$90	$540
Charge Controller	1	$250	$250
Deep-cycle batteries	8	$80	$640
DC to AC Inverter, 2.4 kW, Xantrex DR2412	1	$1,395	$1,395
Back-up generator	1	$1,150	$1,150
Fuses, wiring	1	$150	$150
Labor	16	$50	$800
Whisper-200/1000watt			$2,000
Tower			$1,000
Subtotal			$13,199
US federal renewable tax credit			-$2,000
Total			$11,199

7.33: *Wind/solar hybrid system.*

Life-time savings: $16,879, 145,400 lbs. CO_2, assuming life of 32 years. The expected life is a composite number of the wind system with an expected life of 20 to 25 years, and the PV system rated at 50 years (PV panels have traditionally been rated at 30 years, but this number is likely far too low).

Incremental cost: $7,119. Based on six Sharp 175W PV Modules at $879 each, six mounting brackets at $90 each, one charge controller at $250, eight 6-volt 100 amp-hour deep-cycle batteries at $80 each, one Xantrex DR2412 2.4-kW DC to AC inverter at $1,395, a backup generator at $1,150, fuses and wiring at $150, a Whisper 200 (1000-watt rated output) wind turbine at $2,000, a tower at $1,000, and labor 16 hours at $50 each. Includes cost of replacing the initial deep-cycle batteries after five years with higher-quality deep-cycle batteries that should last you about 20 years, at a cost of $1,280, and additional battery replacements of $640 after 25 years. Subtract the cost of running a power line to your house and installing a transformer, assuming $6,000 for a quarter-mile distance to the nearest power tie-in, and a $2,000 federal tax credit. **New cost**: $13,199, plus $1,280 after 5 years, and $640 after 25 years for battery replacements (total $15,119).

Payback incremental: 13.5 years. **IRR**: 6.8 %. **CROI**: 20.4 lbs./$.

"SELF-BOOTING", SMALL OFF-GRID, RURAL SOLAR ELECTRICITY (PV) INSTALLATION

If price is your first concern, you could try a small, "self-booting" rural solar installation: This system would cost less than a typical rural connection to the

Carbon Buster: 58.8%; $12,590.50

Item	Number	Cost/unit	Output
Sharp 175W PV Module	2	$879	$1,758
Mounting brackets	2	$90	$180
Charge Controller	1	$250	$250
Deep-cycle batteries	4	$80	$320
DC to AC Inverter, 2.4 kW, Xantrex DR2412	1	$1,395	$1,395
Back-up generator	1	$1,150	$1,150
Fuses, wiring	1	$150	$150
Labor	12	$50	$600
Subtotal			$5,803
US federal renewable tax credit			-$1,741
Total			$4,062

7.34: *Small off-grid rural PV system.*

power grid, but would not supply all your electricity needs (thus, the need for a generator). However, your annual savings in power and fixed utility connection costs would allow you to slowly expand your system over the years (thus the term, "self-booting"). The payback for this system is 0 years, though at the loss of some comfort, assuming a $5,000 cost if you had connected your house to the grid. Nominal power output: 350 watts.

Educational Impact of Renewable Energy

One of the positive aspects of renewable energy that is often overlooked is the educational feedback mechanism of renewable systems; when using the utility infrastructure, energy seems to flow from a limitless supply. As soon as you connect your own renewable system (especially if you are living off-grid), you immediately become more aware of the limited nature of your supply, thereby encouraging conservation. For example, having lived off the grid for 16 years has taught me much about power conservation that I would have been happily unaware of in a grid-connected house.

For many, renewable power systems provide a strong incentive to become energy-efficient conservers. It is no coincidence that off-grid users of renewable energy are often the efficiency pioneers, exploring and creating tomorrow's sustainable energy solutions today.

Cool Solar Toys ... that is, ... Tools
SOLAR BACKPACKS

Solar backpacks can supply you with enough power to run your cell phone, iPod or GPS while on the move. They do not supply enough energy to directly run a laptop.

Reware has larger solar panels that supply 6 watts, and they are working on a battery system that would allow you to accumulate enough power to run a laptop, at least for a while. Voltaic backpacks have smaller panels (4 watts), but come with an extensive line of accessories and universal plug-ins that work with virtually any cell phone (iPod connectors optional).

Access: rewarestore.com, voltaicsystems.com

 Carbon Miser: 31.1%; $10,659.90

FLEXIBLE SOLAR PANELS

You can also get rollable solar panels. Extremely lightweight, the larger of these models, in contrast to solar backpacks, *can* be used to run a laptop.

Access: PowerFilm Rollable Solar Module realgoods.com/renew.

Available in 5-watt (0.6 lbs., $119), 10-watt (1 lb., $229), and 20-watt (1.9 lbs., $399) units.

Hydrogen Fuel Cells

Hydrogen fuel cells may well be the way of the future, providing clean energy, assuming we find a way to produce the hydrogen sustainably, and are able to store it effectively (1 to 3 percent of hydrogen stored as a compressed gas in a tank is lost daily through leakage, since hydrogen is the smallest element in the universe).

At the present time, fuel cells are still very expensive, at around $5,000 for a device that produces 1 kW. Ballard Power, a producer of fuel cells, is testing residential applications in the Japanese market. Residential use of hydrogen fuel cells is still extremely rare in the US and Canada.

8

Green Heating

Solar Design

YOU ARE PROBABLY ALREADY MAKING USE of green heating energy without being aware of it: a typical house can derive 30 percent of its annual heating requirements from passive solar heating (the sun coming in through your windows).

Good passive solar design can have a huge impact on your heating (and cooling) bills. And some of the best solar design features you can have are absolutely free, such as orienting your house along an East-West axis, and maximizing south-facing windows (Fig. 8.1).

Currently, city developments are generally designed with absolutely no regard to solar access. Yet, with a single stroke of the pen, a city developer can have a major impact on that subdivision's future energy bills and environmental impact by siting the houses along an East-West axis, providing solar access for all, at no extra cost to the city, or house purchasers; solar access is not yet enshrined as a basic right in the US Constitution or Canada's Charter of Rights, but some day it may be.

What can you do to increase the passive solar effectiveness of your existing home?

8.1: *The home of architect Douglas Cardinal features extensive thermal mass to even temperature fluctuations, and a large window area facing south to capture winter sun.*

DOUGLAS J. CARDINAL ARCHITECT LTD.

Weatherproofing

One of the least exciting, and most low-tech, applications of efficiency technologies is also one of the most effective ones: we are talking here of weatherstripping.

8.2: *This house features double insulation (R-40), airtight design, and heat-recovery ventilation (HRV), all cornerstones of efficient passive solar design.*

Indeed, according to the Rocky Mountain Institute, the humble caulking gun, duct tape, and other simple technologies now provide the US with two-fifths more energy than does the entire domestic oil industry.

Weatherproofing has excellent paybacks, and can reduce your heating bills by up to 20 percent. (In older homes air leakage represents 25 to 40 percent of heat loss.)[94] If you add up all the cracks and gaps in a typical house, there is enough for a medium-size dog to walk through. While air exchange is vital, too much air infiltration provides significant loss of warmed air, with no significant health gain. In older homes, you do not have to be too concerned about getting too airtight, as it is very difficult.

If your home is more recent, and airtight, install a heat recovery ventilation (HRV) system. The system ensures that you receive controlled amounts of fresh air, preventing the "sick building syndrome" that some super-efficient houses displayed in the 1970s due to lack of air circulation, yet allows you to recover a significant portion of your waste heat to keep the house warm (Fig. 8.2).

Carbon Buster/Miser Recommendation: Seal air leaks in your house.

5-year savings: $855, 6,600 lbs. CO_2, 55,000 cu. ft. natural gas.

Life-time savings: $1,369, 10,600 lbs. CO_2, assuming life of 8 years.

New cost: $ 230; including $150 for materials and eight hours of your time at $10/hour.

Payback new: 1.3 years. IRR: 73.5 %. CROI: 46.1 lbs./$.

Insulation

Most houses in North America have inadequate insulation – better insulation is one of the key elements of successful passive solar design.

If your attic is easily accessible, adding more insulation has excellent paybacks; pile it on!

Carbon Buster/Miser Recommendation: Add R-40 cellulose insulation to unheated attic.

5-year savings: $438, 3,400 lbs. CO_2, 28,100 cu. ft. of natural gas.

Life-time savings: $7,000, 54,200 lbs. CO_2, assuming life of 80 years.

 Carbon Miser: 34.6%; $11,952.70

New cost: $564; $504 for materials, $60 for six hours of your time. At an insulation level of R-40, you should be able to get free rental of the blower thrown in.

Payback new: 6.4 years. IRR: 15.5 %. CROI: 96.1 lbs./$.

Note: It is important to wear proper breathing gear to protect you from inhaled particles when installing fiberglass or cellulose insulation. Use of a blower is recommended when installing cellulose to prevent clumping. Take care that ventilation is not impeded by newly installed insulation, to prevent ice buildup or moisture problems.

In other areas of the house, especially the walls, best paybacks are achieved during construction, when the cost of adding insulation is often not much more than the cost of the insulator itself. A lot more effort (and expense) is required to retrofit your home with better insulation once it is built.

However, you can increase your paybacks by insulating at the same time when you replace your furnace and/or air conditioner. Better insulation will reduce your heating and cooling loads, allowing you to save hundreds of dollars just from the ability to purchase smaller-capacity cooling and heating equipment. In addition, the smaller-sized equipment will run more efficiently, adding another few hundred dollars to your savings.

The basement, if uninsulated, is another prime target for cost-effective remediation: if the rest of the house is reasonably well insulated, heat losses from the basement can account for up to 20 percent of your heating bill.[95]

Access: To find a certified home energy auditor or contractor, see energystar.gov/index.cfm?c=home_improvement.hm_audits and resnet.org.

Carbon Buster Recommendation: Add R-12 insulation to basement walls.

5-year savings: $367, 2,800 lbs. CO_2, 23,500 cu. ft. of natural gas.

Life-time savings: $5,865, 45,400 lbs. CO_2, assuming life of 80 years.

New cost: $1,005; $855 for materials (2-by-4 studs, R-12 fiberglass bats, vapor barrier, glue and nails), $150 for 15 hours of your time.

Payback new: 13.7 years. IRR: 7.3 %. CROI: 45.2 lbs./$.

Windows

High-Efficiency, Windows

Inch for inch, the average window loses ten times as much heat (or coolth) as the average wall. Yet, just as with insulation, high-efficiency windows have the best payback during construction. Once the house is built, it takes a very long time to recover the cost of new windows from energy savings alone.

However, if you need to replace windows anyway, and maybe your air conditioner or furnace, this is the time to pay extra for superior insulating value; it may allow you to downsize the A/C unit or furnace as well.

Try to "tune" your windows, to maximize energy efficiency and comfort based on the windows' orientation:

Type of Window	Window R-value (in hr ft² °F / Btu) (higher is better)	Window U-value (in Btu / hr ft² °F) (lower is better)
Single pane	1.0	1.0
Double pane	2.0	0.5
Double pane +argon, low-e coating	2.5	0.4
Superwindow - double pane, two low-e films, two argon-filled spaces	5.0	0.2
Advanced Superwindow - e.g., double pane, three low-e films, krypton-filled	15.0	0.07

8.3: *Insulation effectiveness (R-value and U-value) of select window types, whole window performance (not center-of-glass).*

North-facing window: high U-factor, to maximize insulation, high visible transmittance, to maximize visible light gain (see Figure 8.3 for R- and U-value conversions).

East- and west-facing windows: low solar heat-gain coefficient to minimize summer heating, when the sun's rays are horizontal during sunrise and sunset.

South-facing windows: high solar heat-gain coefficient, to maximize solar gain in winter, if you live in an area with winter-heating. Install an overhang or awning over south-facing windows to minimize summer heat-gain. This will not interfere with winter heat gain, since the angle of incidence of the sun is low, but will greatly reduce unwanted summer heat.

It is a good idea for all windows to have infrared-reflective coatings, which keep heat out in summer, and heat in during winter.

Finding manufacturers making windows over R-5 is not easy. Reported extra costs for the windows range from 10 to 15 percent, to $1,000 dollars per window. This is not an established market, and your negotiation skills may determine the price more than anything else.

US: alpeninc.com, southwall.com (R-12). Canada: duxtonwindows.com (R-14, 5 pane)

Window Retrofits

Rather than replacing your windows, provided they are in reasonably good shape, it is cheaper to retrofit them for higher insulation value.

Retrofitting your windows with a layer of acrylic may cost you around $25/sq. ft. (installed). It would take a while to pay for the retrofit out of the savings.

Window Kits

Window kits, available at hardware stores, are sets of shrinkfoil with double sided tape. You apply the double-sided tape to the window frame, apply the foil, and tighten it by heating the foil with a hair-dryer. The resultant film is crystal clear, and can be cleaned with a sponge (though it *is* susceptible to sharp objects).

 Carbon Miser: 34.6%; $11,952.70

The window kit will save you the equivalent of 2.5 gallons of heating oil per square yard per year if you have double-paned windows, five gallons on single-pane windows, in addition to saving you money on cooling energy.

Window kits that are mounted so as to get a lot of direct sun exposure may only last a single heating season. Otherwise, they can last many years. For example, a window kit I installed on our east side 13 years ago, additionally protected by an overhang, is still going strong. That little square yard of plastic film has saved us nearly $100 of energy in the last decade. Not bad for $1 of window kit and five minutes of my time.

New windows that open by means of a crank can usually receive the window kit as a permanent installation. In older slider or double-hung windows, it may only be possible to apply the kit for the winter season, if at all.

 Carbon Buster/Miser Recommendation: Install window kits ("shrink foil") on 50 percent of your windows.

5-year savings: $428, 3,300 lbs. CO_2, 27,500 cu. ft. of natural gas.

Life-time savings: $171, 1,300 lbs. CO_2, assuming life of two years.

New cost: $147; $23 for materials to cover 18.5 square yards of window, assuming 25 percent of foil is wastage, and four hours of your time at $10 per hour. You will have to apply the window kit 2.5 times over five years.

Payback new: 0.7 years. **IRR:** 101.9%. **CROI:** 20.9 lbs./$.

Energy-Efficient Landscaping

Even landscaping can affect your heating (and cooling) bill: planting trees, especially evergreens, on the north-side of your home can reduce your heating bill by 10 percent or more by reducing cold wind reaching your house and removing heat through convection (while raising your property value by the same percentage). Generally, the lee of the trees is effective for a distance of up to three times the height of the trees.

Your Furnace
High-Efficiency Furnaces

Today's oil and gas furnaces and boilers have advanced rapidly in efficiency. While pre-1992 furnaces and boilers typically had an efficiency of 55 to 65 percent, today's best condensing furnaces reach efficiencies of 99 percent (Figure 8.4).[96]

However, even if you do not change your furnace, you can achieve major savings by sealing gaps in your air ducts and insulating them, especially where they pass

Furnace	Gas Boiler	Oil Boiler
Pre-1992	55-65%	55-65%
1992	78%	80%
Energy Star 2006	90+%	85+%
Condensing	94-99%	84-99%
Sources: E Source, ACEEE, Energy Star		

8.4: *Efficiency of furnaces and boilers (higher is better).*

Carbon Buster: 64.5%; $14,677.60

through unconditioned space. This applies both to heating and cooling, if you air-condition your house via the furnace ducts.

Furnace Maintenance

Another way to increase the efficiency of your furnace is to get a tune-up for it. A Denver, Colorado study found that a $150 tune-up resulted in 12 percent savings of heating costs. In 50 to 80 percent of the cases, furnaces benefited from fan adjustments, lowered rise heat, taping of ducts or plenum, resetting of anticipators, cleaning and oiling of blowers, and replacements of dirty or clogged filters. In 17 percent of cases some ducts even had to be reconnected.[97]

The payback for this measure is excellent, at 1.9 years.

 Carbon Buster/Miser Recommendation: Get a tune-up for your furnace.
5-year savings: $385, 3,000 lbs. CO_2, 24,700 cu. ft. of natural gas.
Life-time savings: as above assuming life of five years.
 New cost: $150 parts and labor.
Payback new: 1.9 years. **IRR:** 42.6%. **CROI:** 19.9 lbs./$.

Motorized Combustion Air Dampers

Motorized combustion air dampers prevent the loss of hot air from uncontrolled fresh air access in your furnace room, saving you heating fuel.

Access: hoyme.com

 Carbon Buster Recommendation: Add motorized combustion air dampers to fresh-air intake for the furnace.
5-year savings: $171, 1,300 lbs. CO_2, 11,000 cu. ft. of natural gas.
Life-time savings: $684, 5,300 lbs. CO_2, assuming life of 20 years.
New cost: $426; $146 materials and $280 labor.
Payback new: 12.5 years. **IRR:** 5.0 %. **CROI:** 12.4 lbs./$.

Thermostats

Setting back the temperature of your house when you are not home, or when you are sleeping, saves about 1 to 3 percent of heating costs per degree of setback.[98] Mechanical thermostats are common in households. Electronic or programmable thermostats have the added benefit of allowing you to time your heater so that the house gets warmed 15 minutes before your wake-up time, and 15 minutes before you come home from work.

Simple programmable thermostats are $40, fancy ones (that allow you to create different settings for the weekend) up to $150.

Fireplaces

Fireplaces — pretty to look at — tend to be a net energy drain on your home, accounting for 14 percent of heating losses in one study. In addition to major

 Carbon Miser: 36.8%; $12,765.30

flue losses due to leakage when not in use, the fire draws hot air up the chimney which gets replaced by cold air infiltrating from the outside.

Three things you can do to increase fireplace efficiency:

- Reduce leakage. Seal cracks with heat-resistant sealant, and make and use a removable flue plug when the fireplace is not in use.
- Install a motorized or manual chimney cap that closes off the chimney when not in use to reduce hot air loss.
- Replace the fireplace with a high-efficiency wood stove.

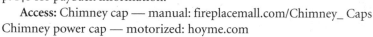

Energy myths

It takes more energy to heat a house back up after letting it cool down, than leaving it warm in the first place. Not so. Heat loss is directly proportional to temperature differential between inside and outside. The lower the setback temperature, the higher the savings will be.

For natural gas fireplaces, turning off the pilot light during the summer will create additional savings.

Carbon Buster/Miser Recommendation: Install a manual chimney cap to close off the chimney when not in use to reduce hot air loss. See p. 140 for payback information.

Access: Chimney cap — manual: fireplacemall.com/Chimney_ Caps Chimney power cap — motorized: hoyme.com

Wood Stoves

Wood stoves can significantly reduce your carbon output and your heating costs. The impact on your pocketbook and the environment depends on a number of factors:

1. Do you have access to a cheap source of wood?
2. Is the wood you are using from a sustainable source?
3. Are you using a high-efficiency stove that offers virtually complete combustion?

Check out epa.gov for information on EPA certified wood stoves.

Pellet stoves are another way to convert a waste product (sawdust) into a valuable heating fuel. Pellet stoves can be very cost-effective, as well as convenient: many pellet stoves have an electrical hopper that feeds pellets into the combustion chamber automatically. As oil and natural gas prices are soaring, the pellet stove industry is growing rapidly.

Geothermal

The original use of the term geothermal derives from the use of ground heat in areas where faults in the earth's surface crust bring magma or hot water close to

Carbon Buster: 67.4%; $15,768.20

the surface. For example, Hawaii and Iceland make extensive use of this kind of geothermal heating.

In more recent times, geothermal has often come to mean the use of a (ground-source) heat pump connected to an underground pipe system that circulates a liquid, i.e., water or antifreeze, through the ground, or even a lake. Some people prefer to refer to this system as "earth energy" instead. Rather than relying on unique geological conditions, this method simply concentrates the environmental heat found outside into your home, in a process that is the reverse of that used in your refrigerator.

Geothermal systems have the advantage that they can also be used to air-condition your home in the summer, in which case the process is *exactly* like that used in your fridge.

Unlike furnaces, which currently have a maximum efficiency of 99 percent, heat pumps have a theoretical efficiency of 300 percent, which means that for each unit of energy you put into the system you derive three units back. This is of course only possible in an open system, where you derive the extra energy from your environment, to satisfy the second law of thermodynamics.

However, heat pumps require electricity to operate, and large amounts of it. The eco-efficiency of a heat pump system therefore largely depends on your source of power.

The payback of your system depends on the quality of your soil (dense soil is cheaper), whether you have enough land to install the system horizontally or need to install vertically (horizontally is cheaper), and whether you can use water — an open-loop system — or must use antifreeze as a heat transfer medium — a closed loop system. (Open-loop is cheaper.)

> ### Energy Quotes
>
> Figures indicate that more than 95 percent of all geothermal heat pump owners would recommend a similar system to their friends and family.
> — California Energy Commission[99]

For an additional $1,000 you can add a Desuperheater, which allows you to use your geothermal system to supply domestic hot water, in addition to space heating and air conditioning. An added bonus of the Desuperheater is that your domestic hot water in the summer is free of cost, as the heat pump system heats the water by using the heat extracted from your house for cooling.

Recommendation: Replace furnace and air conditioning system with ground-source heat pump (geothermal), add Desuperheater to replace hot water tank.

5-year savings: $4,292, 24,388 lbs. CO_2, 386,000 cu. ft. of natural gas, but requires the use of an additional 17,521 kWh of electric power. Your power bill in this scenario will rise from $275 per year to $622, but your annual natural gas costs of $1,205 for heating and hot water will be eliminated. (Note: the additional 4,400 pounds of CO_2 emitted per year due to higher power consumption have been subtracted from the natural gas savings in this calculation to yield net CO_2 savings.)

 Carbon Miser: 38.3%; $13,299.90

Life-time savings: $17,170, 97,551 lbs. CO_2, assuming life of 20 years.[100]

Incremental cost: $5,873, saving $2,627 in furnace costs.

New cost: $7,500 materials and labor, for a 3 tonne unit, plus $1,000 for a Desuperheater, which allows you to produce domestic hot water with the heat pump.

Payback incremental: 6.8 years. IRR: 13.4 %. CROI: 16.6 lbs./$.

Payback new: 9.9 years. IRR: 7.9 %. CROI: 11.5 lbs./$.

Note: if you install all other Carbon Buster recommendations first, the savings potential of geothermal drops from $4,292 per five years to $2,212 per five years, and the payback changes from 9.9 years to 19.2 years for new systems. See p. 140 for details.

You can receive a $300 US federal tax credit for any geothermal heat pump that meets the following efficiency requirements: EER (energy efficiency rating) 14.1, COP (coefficient of performance) 3.3, for a closed- loop system, EER 16.2, COP 3.6 for an open-loop system, EER 15, COP 3.5 for a direct expansion system, or for an electric heat pump water heater that has an Energy Factor of 2.0.

For updates on incentives for geothermal systems, especially state ones, check geoexchange.org/incentives/incentives.htm.

Domestic Hot Water
Solar Collectors

Solar collectors can be used to supply a substantial portion of your family's hot water needs, anywhere in the US or Canada. Solar collectors convert a far larger portion of sunlight (about 60 to 87 percent) to useful energy than photovoltaic panels (typically 16-18 percent). Your home, or a location on your land not too far from your home, must have fairly unobstructed access to sunshine, particularly from 10 am to 2 pm, the time of peak solar gain.

Canadian Government Subsidies: nrcan.gc.ca; energytaxincentives.org. For US incentives see p. 100

Carbon Buster Recommendation: Install a thermal solar collector for your hot water needs.

5-year savings: $1,216, 9,400 lbs. CO_2, 78,000 cu. ft. of natural gas.

Life-time savings: $4,866, 37,700 lbs. CO_2, assuming life of 20 years.

New cost: $3,715; $4,307 for materials, $1,000 for labor, minus a 30 percent federal tax credit of $1,592. (Up to a maximum of $2,000 for photovoltaics and solar water heating.)[101]

Payback new: 15.3 years. IRR: 2.7 %. CROI: 10.1 lbs./$.

Tankless, On-demand Water Heaters

On-demand water heaters, unlike regular water heaters, do not store hot water in a tank. Instead, they only start heating the water as it is required, i.e., "on demand." The advantage of this system is that you do not experience the stand-by losses associated with a tank water heater, allowing you to save 40 percent of your water heating cost.

Another advantage is that you never run out of hot water, as more hot water can be produced instantaneously at any time — no more fighting over the shower after a long trip.

However, unlike tank heaters, there is a limit to the amount of hot water that can be delivered at the same time. If this may be an issue in your house (e.g.: two showers and a washing machine running at the same time), size your on-demand water heater accordingly, or install two heaters in series.

When buying a tankless water heater, insist on an electronic ignition model. If your instantaneous water heater had a pilot light, you would be negating a large portion of your savings. In my Bosch water heater, the two long-life D-cell Varta batteries only have to be replaced once every six years (by contrast, a pilot light would have cost me between $240 and $720 in natural gas over the same time, according to utility figures).[102] In fact, some of the newer models have a micro-turbine in the water supply line, generating the electricity required to start the flame from the water flow created when the tap is opened. Very cool!

Carbon Buster/Miser Recommendation: Buy electronic-ignition tankless water heater.

5-year savings: $614, 4,750 lbs. CO_2, 39,000 cu. ft. of natural gas.

Life-time savings: $1,350, 10,453 lbs. CO_2, assuming life of 11 years.

Incremental cost: $273; $578 for tankless water heater minus cost of regular water heater at $305.

New cost: $578 materials and $150 installation.

Payback incremental: 2.2 years. **IRR:** 44.2 %. **CROI:** 38.3 lbs./$.

Payback new: 5.9 years. **IRR:** 12.0 %. **CROI:** 14.4 lbs./$.

Tank Water Heaters

An investment in heater insulation is a wise and inexpensive move. If you decide *not* to spring for a tankless water heater, for about 50 cents you can buy a 3-foot foam section of sleeve insulation for ½" hot water pipes at a hardware store. A slit on the side allows you to slip these easily over the hot water pipes coming out of your water tank. Insulating the first 9 feet gives you the highest pay-backs.[103] An insulating jacket (also available in hardware stores) placed over the tank leads to additional savings.

Recommendation: Insulate first 9 feet of water tank hot water pipe.

5-year savings: $18, 140 lbs. CO_2, 1,200 cu. ft. of natural gas.

Life-time savings: $72, 560 lbs. CO_2, assuming life of 20 years.

New cost: $1.50 for three 3-foot sleeves.

Payback new: 0.4 years. **IRR:** 240 %. **CROI:** 371.5 lbs./$.

 Carbon Miser: 39.9%; $13,913.70

Cooling

WITH A TYPICAL ANNUAL POWER CONSUMPTION of 2,784 kWh, air conditioning can be the single largest user of electricity in the home, especially in houses in warmer regions.

Depending on where you live, good passive design can reduce or eliminate altogether the need for air conditioning in the first place. Good thermal insulation and high mass inside the building will even out temperature fluctuations. The revival of cob (a mixture of sand, straw, and clay not unlike adobe) makes use of this thermal mass in new houses. (See *The Hand-Sculpted House: A Philosophical and Practical Guide to Building a Cob Cottage*, Appendix M.)

What are good cooling strategies for existing homes?

Green Coolth — Passive Solar Design

Many of the principles used to conserve heat (see "Green Heating," p. 107) work equally well to keep heat out in summer.

The following techniques, discussed previously, are as effective in reducing cooling loads as they are in reducing heating loads:

1. Add insulation to attic and basement, p. 108
2. Install window kits, p. 110
3. Choose high-efficiency windows when replacing old ones, p. 109
4. Energy-efficient lighting will cut your cooling needs by one-third of the lighting energy saved, p. 56

Cheap Cooling Strategies

In addition, there are a number of ways to reduce cooling loads that may not cost much:

- Use white or light gray roofing material

A roof with a high albedo (reflectance) will greatly reduce cooling loads (in fact, it also reduces heating costs in cooler regions, since light-colored roofs delay the melting of snow, adding the snow itself to your roof-insulation).

• Avoid using tinted glass that cuts too much visible light

Some tinted glass may sufficiently lower the light levels in your rooms that you need to use artificial lighting, which in turn increases the cooling load (expect each kWh of power consumption to result in about one-third of a kWh in cooling consumption).

However, spectrally selective tinted or coated glazing can help reduce solar gain while providing nearly as much visible light as clear glass.

• Avoid the reverse stack effect in cooled buildings

Cold air will flow out at the lowest points, and draw in warm air at the top. Therefore, providing good door and window seals, and not leaving outside doors and windows open, can reduce cooling costs.

• Energy-efficient landscaping

Planting deciduous shade trees on the east, south or west sides provides cooling in summer, yet allows most of the sun to enter during the heating season. Cooling savings of 10 percent and more have been reported from shaded homes. However, you will have to be a little careful around solar PV systems: while energy loss is directly proportional to shading with solar thermal collectors, small amounts of shade can lead to disproportionately large losses in electrical gain with photovoltaic systems, even if only a small portion of the array is affected. Ensure that the photovoltaic panels have full access to sunlight, especially from 10 am to 2 pm.

A tree can drop the temperature by about 1.2 to 2.3 °F

> **Quick Fact**
>
> The net cooling effect of a young, healthy tree is equivalent to ten room-size air conditioners operating 20 hours a day.
> — US Department of Agriculture

The National Wildlife Federation estimates that there are between 60 to 200 million spaces along our city streets where trees could be planted, which would absorb 33 million more tons of CO_2 and save $4 billion in energy costs.

Planting creeping vines on exposed house walls also provides effective reductions in cooling requirements.

Technological Solutions
Ground-Source Heat Pumps (Geothermal)

Ground-source heat pumps are even more efficient for cooling (and heating) than air-to-air heat pumps. The reason for this is that the temperature of the ground a few feet below the surface is relatively stable (typically 39 - 54°F, depending on where you live). Therefore, you can usually derive more coolth from the ground in summer, and more heat in winter, than you could from an air-to-air unit. See page 113 for calculations of geothermal heat pump paybacks.

Central Air Conditioning

Central air-conditioning systems, while not as efficient as ground-source heat pumps, are the most common cooling systems installed in new homes today.

Check energystar.gov for the latest Energy Star-rated systems.

Room Air Conditioners

Room-sized air conditioners are less efficient for a given Btu rating than central air conditioners. However, if your cooling needs are modest, or if your home is not very large, room air conditioners will actually be *more* efficient. Check Figures 9.1 and 9.2 for the top Energy Star rated air conditioners.

Brand	Model	Louvered Sides	Casement	Capacity (BTU/hr)	Energy Efficiency Ratio (EER) (higher is better)	Federal Standard (EER)	Percent Better	Purchase Cost ($)
Friedrich	XQ06J10A	No		6300	11.5	9.0	0.28	$400
Friedrich	XQ06L10*	Yes	None	6300	11.5	9.7	0.19	$346 to $455
Friedrich	SS09L10*	Yes	None	9100	11.5	9.8	0.17	
Friedrich	SS08L10*	Yes	None	8400	11.4	9.8	0.16	$577 to $669
Carrier	ACA051T	Yes	None	5400	11.2	9.7	0.15	$200
Frigidaire	FAA056M7A	Yes	None	5400	11.2	9.7	0.15	
Kenmore	72059, 75052	Yes	None	5700	11.2	9.7	0.15	$180
Sharp	AF-P50CX, AF-S50CX, AF-S50DX, AF-S55CX	Yes	None	5000	11.0	9.7	0.13	$200 to $258
Carrier	ACA051B	Yes	None	5200	11.0	9.7	0.13	$197
Frigidaire	FAA053M7A	Yes	None	5200	11.0	9.7	0.13	$144 to $160
Frigidaire	FAA053P7A	Yes	None	5200	11.0	9.7	0.13	
Frigidaire	FAA055M7A	Yes	None	5200	11.0	9.7	0.13	$185
Frigidaire	FAA055N7A	Yes	None	5200	11.0	9.7	0.13	$150
Frigidaire	FAA055P7A	Yes	None	5200	11.0	9.7	0.13	$150 to $199
Frigidaire	FAA056P7A	Yes	None	5200	11.0	9.7	0.13	
White-Westinghouse	WAA053M7A	Yes	None	5200	11.0	9.7	0.13	
Friedrich	XQ05J10A	No		5500	11.0	9.0	0.22	$396
Friedrich	XQ05L10*	Yes	None	5500	11.0	9.7	0.13	$360 to $430
Airtemp	B7X05F2A	Yes	None	5600	11.0	9.7	0.13	$139
Carrier	GCA061T	Yes	None	5600	11.0	9.7	0.13	
Kenmore	72055	Yes	None	5600	11.0	9.7	0.13	$220

GODO STOYKE/CARBON BUSTERS

9.1: *Energy Efficiency Ratio (EER) of most efficient Energy Star room air conditioners, 5,000 to 9,900 BTU/hr (higher EER is better. Source: Energy Star).*

Carbon Buster: 72.4%; $17,598.40

Quick Fact

There are about 60 to 200 million spaces along our city streets where trees could be planted. This translates to the potential to absorb 22 million more tons of CO_2 every year, and saving $4 billion in energy costs.

— National Wildlife Federation

Sizing Air Conditioners

Sizing air conditioners, regardless of system, is critical. Over-sized systems are actually *less* efficient than properly sized ones. One of the reasons for this is that larger systems will reach the set temperature faster — too quickly, in fact, to remove enough of the room's moisture. Therefore, you will feel less comfortable, and require more air conditioning.

Brand	Model	Capacity (BTU/hr)	Energy Efficiency Ratio (EER) (higher is better)	Purchase Cost ($)
Friedrich	SS10L10*	10,400	12.0	$625 to $744
Friedrich	RS12L10, SS12L10*	11,800	11.8	$538 to $779
Friedrich	RS10L10, SS10J10AR	10,200	11.7	$523 to $699
Friedrich	KS10L10*	10,000	11.5	$516
Friedrich	SS12L30*	12,100	11.5	$712 to $859
Friedrich	RS15J10	14,500	11.1	$595
Friedrich	RS10J10	10,000	11.0	$500 to $524
LG Electronics	LB1000ER, LW1004ER, LW100CS	10,000	11.0	$303
Sharp	AF-S100DX	10,000	11.0	$325
Friedrich	KS12L10*	11,600	11.0	$584
Friedrich	RS12J10A, SS12J10AR	11,750	11.0	$525 to $710
Friedrich	SS12J30D	12,000	11.0	$739
Kenmore	74156	15,100	11.0	
Climette	CA1816AR	17,300	11.0	
Comfort Aire	RADS-181	17,300	11.0	$679
Maytag	M7D18E2A	17,300	11.0	
Friedrich	SM18L30*	17,800	11.0	$829 to $939
General Electric	AGF10AB	10,000	10.9	$319
Friedrich	KS15J10	14,500	10.9	$1537
Airtemp	B7Y10F2A	10,000	10.8	$199 to $238
Amana	AC103E	10,000	10.8	
Amana	ACD105E	10,000	10.8	$250 to $285
Comfort Aire	RADS-101	10,000	10.8	$327
Crosley	CAR10RSL	10,000	10.8	
Crosley	CAR10RSP	10,000	10.8	
Fedders	A7Y10F2A	10,000	10.8	$259 to $375
Fedders	A7Y10F2B	10,000	10.8	$294
Friedrich	CP10A10	10,000	10.8	$275 to $339
Friedrich	CP10A10*	10,000	10.8	$275 to $339

9.2: *Energy Efficiency Ratio (EER) of most efficient Energy Star room air conditioners, 10,000 to 19,000 BTU/hr (higher EER is better. Source: Energy Star).*

 Carbon Miser: 39.9%; $13,913.70

10

Liquid Assets

Where Does All the Water Go?

A CCESS TO HIGH-QUALITY WATER may become one of the most important issues facing us in the 21ˢᵗ century. Water is intimately linked with the resource consumption in your home. Not only does it take energy to collect, treat and distribute water, water use also affects your energy bills through the consumption of hot water. Water conservation therefore provides the opportunity to save twice: on water bills *and* on energy bills.

The single biggest average water user is the yard, followed by toilets and showers (Figure 10.1).

The best way to reduce water use due to clothes washing is through the purchase of a front-loading washer, discussed on page 82.

10.1: *Residential water use (Lawrence Berkeley Laboratory).*

Toilets

Responsible for 32 percent of your water consumption, toilets are the single largest users of water *inside* your home. Replacing your conventional toilet for water savings has a payback of 11 years if your current toilet consumes 3.5 gallons per flush, and five years if your toilet consumes seven gallons per flush. Even higher savings can be achieved using a dual-flush toilet, which has a 1.6 gallon and a 0.8 gallon flush mode — for example, the Caroma Royale (sustainablesolutions.com).

Carbon Buster/Miser Recommendation: Buy two low-flush (1.6-gallon) toilets to replace 3.5-gallon toilets.

5-year savings: $207, 662 lbs. CO_2, 102,000 gallons of drinking water.
Life-time savings: $826, 2,600 lbs. CO_2, assuming life of 20 years.
New cost: $450; $300 for two low-flush toilets using 1.9 gallons each, $150 for installation.
Payback new: 10.9 years. **IRR:** 6.6 %. **CROI:** 5.9 lbs./$.

GABRIEL WONG/CARBON BUSTERS

THAT'S NOT WHAT I MEANT WHEN I SAID WE FLUSH AWAY GALLONS OF DRINKING WATER!

Quick Fact

"Nothing is too good for my toilet bowl."
Old standard water commodes waste seven gallons of extensively processed, high-quality drinking water per flush.

Quick Fact

40 percent of all toilets leak.
— Environment Canada

Just pour a few drops of red food coloring into your toilet tank to check if yours does.

Checking for Leaks

There is an easy and quick method to check if your toilet seal is leaky and needs to be replaced: simply put a few drops of red food coloring into your toilet tank.

Wait five to ten minutes, and check your toilet bowl: if the water in the toilet bowl has turned pink, you need to replace your tank seal.

Composting Toilets

Modern composting toilets work extremely well, and can eliminate one-third of your interior water use. Good composting toilets are not cheap (around $4,000 to $5,000). However, they can be extremely cost-effective if you are not connected to a city sewer, and would need to spend $15,000 on a septic field as an alternative.

Access: Phoenix composting toilet: compostingtoilet.com

Cost: Two people PF-199 $4,200, Four people: PF-200 $4,600, Eight people: PF-201 $5,300

Showers

Showers are the second largest water user in the house after toilets. High-efficiency, low-flow showerheads provide excellent paybacks (around six months), since they save you water as well as heating fuel.
Access: realgoods.com

Baths

Currently, there is no technology to reduce water usage in the bath (other than bathing with a friend).

Yard Water Usage

One way to drastically lower your water consumption is to redesign your yard to naturescaping (check page 125), or through xeriscaping. Native vegetation provides better habitats for native plants and animals. Native plants are also well adapted to your local climate, tend to be much tougher, and generally require far less water. Xeriscaping takes this one step further by providing you with a yard that needs little or no water through the use of landscaping features like rocks or pine cones, and drought-adapted plants.

Carbon Miser: 40.2%; $14,120.30

11

Environmental Goods and Services

T HIS BOOK PRIMARILY EXAMINES THE IMPACT OF ENERGY use on the environment. However, you also have an impact through all your other buying decisions. As a detailed analysis of the carbon footprint of your non-energy purchases is beyond the scope of this publication, we will look briefly at a few areas where you can have a positive impact.

Sustainable Wood

The Forest Stewardship Council (FSC) provides the most stringent certification system for wood that is produced in a responsible, sustainable manner. FSC certification is becoming a minimum standard among many organizations, including printers, furniture makers, and building associations. For example, the use of FSC-certified wood will earn builders a point under the LEED (Leadership in Energy and Environmental Design) certification system of the US and Canada Green Building Councils usgbc.org, cagbc.org. FSC certification or equivalent has also become a requirement of IKEA's wood purchasing policy.[104]

Sustainable development is like teenage sex — everybody claims they are doing it but most people aren't, and those that are, are usually doing it very badly.

— unattributable

Check out the data base on the website of the Forest Stewardship Council fsc-info.org to find FSC-certified wood products in your area.

Food

Currently, 90 percent of our food calories are based on fossil-fuel sources; only 10 percent are derived more directly from the sun.[105]

Organic and vegetarian foods both offer substantial health advantages for their consumers, but also for the planet. For example, low-input sustainable farming increases the carbon-sink capacity of the soil by adding compost and green manure to the ground, and greatly reduces the need for petrochemicals that supply synthetic fertilizers and pesticides in conventional farming.

For a list of organic restaurants and cafes in your area check:

localharvest.org/restaurants

For vegetarian restaurants and cafes see:

happycow.net

vegetarian-restaurants.net

vegdining.com

You can also reduce the energy requirements of your food by patronizing the local farmer's market. In addition to being a fun way to spend a Saturday morning, you support the local economy, get fresh produce and sometimes meat, and reduce the fossil fuel energy required to ship food across the country.

Recycled Products/Recycled Paper

We use 728 pounds of paper per person per year.[106] The use of recycled paper greatly reduces the environmental impact associated with virgin paper use. One metric tonne of recycled paper saves:

- 4,100 kWh of electricity
- 3.2 cubic yards of landfill space
- 360 gallons of water
- 2 barrels of oil
- 7 trees[107]

One paper with a very high recycled paper content is Domtar Sandpiper, manufactured with 100 percent post-consumer waste, processed chlorine free, and FSC certified (domtar.com).

For other recycled product sources check:

recycledproducts.org

recycledofficeproducts.com

store.yahoo.com/greenearthofficesupply

dolphinblue.com

treecycle.com

ecoproducts.com

gracefulearth.com/Coreless

ecomall.com/greenshopping

Recycled-Content Product Directory: ciwmb.ca.gov/RCP

Organic Flowers

Organic flowers won't taste any better, but they will save the environment from thousands of pounds of pesticides, and result in reduced energy requirements (and greenhouse gas emissions) due to reduced manufacture of synthetic fertilizers.

 Carbon Miser: 40.2%; $14,120.30

The first and only on-line organic flower retailer in the US is Organic Bouquet at organicbouquet.com

In Canada you can order from ecoflora.ca

Most of the annual expenses for cut flowers are incurred on just two days a year (Valentine's and Mother's Day), so order in time and avoid the rush.

Protecting Nature at Home

Naturescaping

Naturescaping provides a great learning environment for our children, and provides habitats for native plants and animals. In addition to reducing energy consumption and water use (see pages 111, 118, and 122), naturescaping can reduce stress and provide a joyous escape from the sterility of manicured lawns. Following are some ideas for adding interest and excitement to your yard.

Teaching a child not to step on a caterpillar is as valuable to the child as it is to the caterpillar.

— Bradley Miller

Nesting Aids and Butterfly Flowers

Everyone has heard of bird boxes. But you can also provide boxes for bats, and nesting aids for solitary bees, wasps and bumblebees.

To build a bat box, check out: dnr.state.wi.us and lincstrust.org.uk (The latter is a UK site, but still relevant.)

Or read the book: Merlin D. Tuttle, Mark Kiser, Selena Kiser. *The Bat House Builder's Handbook: 2ⁿᵈ Edition*, Bat Conservation International, 2005. $9.95

For other animal homes take a look at: Bobbe Needham. *Beastly Abodes: Homes for Birds, Bats, Butterflies & Other Backyard Wildlife*, Sterling Publishing, 1996. Available used.

> ## Quick Fact
> "In laboratory research, visual exposure to settings with trees has produced significant recovery from stress within five minutes, as indicated by changes in blood pressure and muscle tension."
> — Dr. Roger S. Ulrich, Texas A&M U.

Green Mortgages

There are a number of energy-efficient mortgages (EEMs) that give you preferred rates, or allow you to add 100 percent of the additional cost of qualifying energy efficiency measures to already approved mortgages. This allows you to take advantage of excellent interest rates for adding energy-efficient features to your new home, or even when just retrofitting your home.

Access: energystar.gov

> ## Quick Fact
> "One acre of forest absorbs six tons of carbon dioxide and puts out four tons of oxygen. This is enough to meet the annual needs of 18 people."
> — US Department of Agriculture

Carbon Buster: 72.6%; $17,805.00

Environmental Investing

You can also look at investing in companies or mutual funds that specialize in sustainable activities. For example:

- Sierra Club sierraclubfunds.com. They offer two funds: Sierra Club Stock Fund and Sierra Club Equity Income Fund.
- Calvert Online calvert.com. To find out how your mutual fund measures up in terms of social responsibility or environment, enter it into the search and find out how the fund performed against Calvert's corporate responsibility criteria.
- Dow Jones Sustainability Indexes sustainability-indexes.com.
- Dow Jones Sustainability Indexes track the financial performance of the leading sustainability-driven companies worldwide.
- Sustainable Business sustainablebusiness.com/progressiveinvestor. Subscribe to receive information on emerging businesses in the sustainable business sector.
- Social Investment Organization socialinvestment.ca. SRI funds in Canada that are widely available for sale to the investing public.
- Canadian Magazine for Responsible Business: corporateknights.ca. You can also download the Responsible Investing Guide at this site.

Putting It All Together

Aᶠᵗᵉʳ ɪᴍᴘʟᴇᴍᴇɴᴛɪɴɢ ᴀʟʟ ᴛʜᴇ ᴄᴀʀʙᴏɴ ʙᴜꜱᴛᴇʀ recommendations for your
family, what have you achieved?

Financial Cost and Gain

Let's take a look at the costs first. As you may have noticed, the Carbon Buster
energy diet is not for the meek. If you want some serious results, you have to do
some serious investing. No financial pain, no financial gain.

The cost of becoming a Carbon Buster is $11,722 at the beginning of the pro-
gram, with an additional $11,640 spent over the next 25 years. That is some
serious money. However, the financial gain is much greater still. Following the
Carbon Buster recommendations will save you $17,805 in the first five years, and
$89,025 over the course of the 25-year program (net savings of $65,665; Fig. 12.1).
This program is not for the overnighter.

If you are the kind of person who would rather invest in that "hot tip" uranium
mine, with a "guaranteed" return of 300 percent over the next 12 months, stick to
installing window kits and sealing air leaks in your home with high, quick returns
(i.e., the top IRR recommenda-
tions of Appendix C, page 136).

However, if you have the per-
sonality that thinks long-term,
in terms of 401(k)s — or RRSPs
in Canada — and mutual funds
rather than miracle stocks, i.e.,
investments where you do not
expect overnight success, but
rather steady, sure, long-term
gains, the Carbon Buster strategy
is right for you.

	Carbon Miser	Carbon Buster
Investment cost program begin ($)	4,484	11,720
Investment cost next 25 years ($)	5,301	11,640
Total cost ($)	9,785	23,360
Savings 5 years ($)	14,120	17,805
Savings 25 years ($)	70,602	89,025
Net savings ($)	60,817	65,665
Carbon emission reductions	40%	73%

12.1: *Costs and savings of Carbon Miser and Carbon Buster
energy efficiency strategies.*

Gᴏᴅᴏ Sᴛᴏʏᴋᴇ/Cᴀʀʙᴏɴ Bᴜꜱᴛᴇʀꜱ

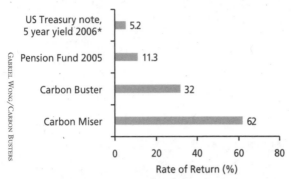

GABRIEL WONG/CARBON BUSTERS

12.2: *Financial returns on some common investment instruments compared to energy efficiency strategies (*yields vary).* [108]

Quick Tip

The Carbon Buster and Carbon Miser strategies provide you with financial returns of 32% and 62% respectively!

As most of us have become painfully aware at the fuel pump, if nowhere else, energy is a component of our family's basic expenses that is growing rapidly in significance.

Investing in energy efficiency is a way to secure the long-term financial stability of your family — every kWh, cubic foot of gas, or gallon of gasoline that your home and car do not need to consume contributes to the future-proofing of your family. And it is a lot safer investment than almost anything else you can get on the investment market (Fig. 12.2).

How about the Carbon Miser strategy?

As a Carbon Miser you spend less money than a Carbon Buster to get the program going: $4,484 at the beginning of the program, and an additional $5,301 over the next 25 years. Your savings are $70,602 over 25 years, or about $20,000 less than for the Carbon Buster.

Though the returns per dollar invested are higher for the Carbon Miser, the net savings over 25 years are slightly lower, at $60,817.

Rising Energy Prices, Falling Efficiency Costs

The financial predictions in this book are based on constant 2005/2006 energy prices and costs of implementing efficiency measures. However, this assumption is conservative.

For the last ten years, two trends have been evident: rapidly rising costs for most types of energy, and falling costs for efficiency technologies.

For example, the cost of a typical compact fluorescent light has dropped from $25 in 1986 to $5 or less in 2006. No one can predict the future with certainty. However, based on past trends it is most likely that the actual savings for Carbon Misers and Carbon Busters will be even greater. In fact, my prediction is that the disparity between energy price increases and efficiency cost drops will be so great that the Carbon Buster strategy will outperform the Carbon Miser one over the next 25 years in total savings even more significantly.

Carbon Miser: 40.2%; $14,120.30

Environmental Gains

Carbon Misers will save more than 3,000 gallons of gasoline over five years, more than 15,200 gallons over 25 years, 192,000 cu. ft. of natural gas over five years and 959,000 over 25 years. They will save 23,600 kWh of electricity over five years and 118,000 kWh over 25 years, more than 138,000 gallons of water over five years and more than 693,000 gallons over 25 years (Fig. 12.3).

Carbon Busters will save more than 3,600 gallons of gasoline over five years, more than 18,200 gallons over 25 years, 313,000 cu. ft. of natural gas over five years and 1,566,000 over 25 years. They will save 24,700 kWh of electricity over

		Carbon Miser		Carbon Buster	
		5 years	25 years	5 years	25 years
Gasoline savings	gal	3,050	15,300	3,650	18,275
	$	8,500	42,600	10,198	50,988
	percent reduction gasoline	44	44	53	53
	CO2 (lbs.)	59,700	298,400	71,508	357,540
Electricity savings	kWh	23,600	118,200	24,744	123,718
	$	2,339	11,694	2,447	12,236
	percent reduction electricity	40	40	42	42
	CO2 (lbs.)	30,000	150,200	31,421	157,104
Natural gas/heating	cu. Ft.	191,800	958,800	313,114	1,565,569
	$	2,988	14,938	4,878	24,392
	percent reduction natural gas	51	51	84	84
	CO2 (lbs.)	23,100	115,600	37,759	188,797
Water	gallons	138,700	693,500	138,700	693,500
	$	282	1,410	282	1,410
	percent reduction water	26	26	26	26
	CO2 (lbs.)	900	4,500	900	4,500
Biodiesel replacing diesel	CO2 (lbs.)	-	-	21,378	106,888
Green power	CO2 (lbs.)	-	-	43,389	216,944
Total	$	14,100	70,600	17,805	89,025
	Percent cost	45	45	57	57
	CO2 (lbs.)	114,500	572,600	206,357	1,031,787
	Percent CO2	40	40	73	73

12.3: *Environmental impacts of Carbon Miser and Carbon Buster energy efficiency strategies (numbers may not add perfectly due to rounding).*

GODO STOYKE/CARBON BUSTERS

Carbon Buster: 72.6%; $17,805.00

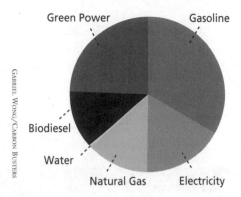

12.4: *Carbon reductions of Carbon Buster energy efficiency strategies (178,512 lbs. of CO_2 over five years).*

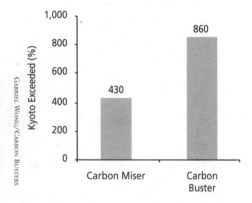

12.5: *Carbon Miser and Carbon Buster energy efficiency strategies relative to fulfillment of Kyoto Protocol (Kyoto target = 100).*

Quick Fact

Leading climate scientists are recommending an immediate reduction in greenhouse gas emissions by 60 to 80 percent. Carbon Busters can actually achieve this goal.

five and 123,700 kWh over 25 years, 138,000 gallons of water over five and more than 693,000 gallons over 25 years).

Carbon Misers will also save 114,000 pounds of CO_2 over the next five years, and 570,000 pounds over the next 25 (Fig. 12.3). Carbon Busters will achieve even higher savings of 207,000 pounds over five and 1 million pounds over 25 years.

Figure 12.4 shows the relative contributions of the various utilities to your carbon reductions with the Carbon Buster approach. You can see that using biodiesel, and subscribing to green power, have a major impact on your emissions. Currently, neither of these measures will save you any money, though this may change in the future, as prices start to reflect environmental costs. I urge you strongly to consider adopting these two measures. The relatively small extra cost of using biodiesel and green power may be among the best 47 cents per day you have ever spent!

The Kyoto Protocol

The aim of the Kyoto Protocol, the first international agreement to take action on greenhouse gases, was signed by the United States, but not ratified. However, numerous US states and cities have picked up the gauntlet and vowed to meet the Kyoto targets. Under the protocol, the US had committed to reduce greenhouse gas emissions by seven percent below 1990 levels, by 2008 to 2012.

Adjusting for increases in residential CO_2 emissions, Carbon Misers will *exceed* the requirements of the Kyoto Protocol in the residential area by 430 percent, and Carbon Busters will exceed Kyoto targets by 860 percent, both at a huge profit, not at a net cost, and years ahead of schedule (Figure 12.5).

Leading climate scientists are recommending an immediate reduction in greenhouse gas emissions by 60 to 80 percent, just to *stabilize* already high, current greenhouse gas levels (in other words, Kyoto is only a baby-step towards the greenhouse gas reductions we have to achieve). Carbon Misers can achieve huge progress towards this goal, with 40 percent carbon emission reductions (37 percent when adjusted for emission increases since 1990). Carbon Busters can actually *achieve* this goal with 73 percent reductions in carbon emissions (67 percent when adjusted for emission increases since 1990; Fig. 12.4).

Share your Success Stories!

I hope this book has inspired you to save energy and contribute to slowing climate change.

The measures discussed in this book are by no means exhaustive, and new technologies are emerging almost daily. Visit the Carbon Buster's Home Energy Handbook website, carbonbusters.org for updates.

We also invite you to share your success stories in implementing efficiency strategies, and any new energy efficiency technology and tricks you have used, at the same website.

You can also contact us via email (handbook@carbonbusters.org) or Fax (1-780-437-1500).

Go to it!

Get updated information and share your energy efficiency success stories at the Carbon Busters website, carbonbusters.org/handbook!

Whatever you can do, or dream you can, begin it. Boldness has genius, power, and magic in it.
— Johann Wolfgang von Goethe

Quick Fact
The Carbon Buster strategy will allow you to exceed the requirements of the Kyoto Protocol by 860%!

Appendices

Appendix A: Top Carbon Buster Measures

Top carbon buster measures, rated by highest carbon dioxide savings in pounds. I=incremental scenario (choose more efficient alternative when replacing worn-out model), N=new scenario (old unit still functional, replace with more efficient model). Negative numbers indicate increased consumption.

Efficiency Measure	CO_2 5 Years lbs. (higher is better)	Incremental $ (lower is better)	Payback Years (lower is better)	Your Savings (higher is better)
I: Honda Insight hybrid car (manual) as primary car (2 seats max.), p. 39	49,867	3,000	2.1	_____
N: Subscribe to green power (after you got geothermal heating/cooling/ water heating), p. 95	47,960	58	N/A	_____
I: Toyota Prius hybrid car as primary family car (5 seats), p. 42	47,139	1,425	1.1	_____
N: Subscribe to green power, p. 95	43,389	53	N/A	_____
N: Subscribe to green power (geothermal, no fuel switching), p. 95	41,030	50	N/A	_____
I: Toyota Echo (manual) as primary family car (5 seats), p. 47	35,319	0	0.0	_____
I: Honda Insight hybrid car (manual) as secondary car (2 seats max.), p. 39	26,793	3,000	3.9	_____
I: Honda Insight hybrid car (auto) as secondary car (2 seats max.), p. 39	24,452	1,550	2.2	_____
N: Geothermal system for heating, AC, hot water, p. 113	24,388	8,500	9.9	_____
I: Geothermal system for heating, AC, hot water, p. 113	24,388	5,873	6.8	_____
I: Mercedes Smart Diesel car as secondary car (2 seats), p. 46	24,369	750	1.1	_____
I: Off-grid, rural solar/wind hybrid power installation, p. 102	22,749	7,119	13.5	_____
N: Use 50% biodiesel for Smart or other diesel, p. 49	21,378	594	N/A	_____
N: Install ground-source heat pump (geothermal), p. 113	19,439	7,500	11.4	_____
I: Replace furnace and air conditioning with ground-source heat pump (geothermal), incremental cost, p. 113	19,439	4,873	7.4	_____
N: Convert primary car to compressed natural gas, p. 49	18,588	2,000	3.0	_____
N: Geothermal system for heating, AC, hot water, after insulation, window kits etc., p. 113	15,948	8,500	19.2	_____
I: Off-grid, rural wind power installation, p. 102	14,260	1,185	3.0	_____
I: Toyota Echo (manual) as secondary family car (5 seats), p. 47	12,939	0	0.0	_____
N: Install a thermal solar collector for your hot water needs, p. 115	9,415	3,715	15.3	_____
I: City-based, grid-connected solar electricity (PV) installation, p. 101	8,000	4,415	35.4	_____
I: Off-grid, rural solar electricity (PV) installation, p. 101	7,022	1,999	7.2	_____
I: Replace top-loading washer with front-loading model (water is heated electrically), p. 82	7,022	53	0.4	_____
N: Eliminate 90% of your power vampires, p. 67	6,733	125	1.2	_____
N: Use cold/cold instead of hot/hot wash mode (electrically heated water) p. 82	6,636	0	0.0	_____

N: Seal air leaks in your house, p. 107	6,621	230	1.3 _____
N: Upgrade medium-efficiency to high-efficiency furnace, no other improvements, p. 111	6,290	3,667	22.6 _____
I: Choose high-efficiency furnace instead of medium-efficiency when replacing, no other upgrades implemented, p. 111	6,290	1,040	6.4 _____
N: Upgrade medium-efficiency to high-efficiency furnace, all other improvements, p. 111	5,281	3,667	26.9 _____
N: Use warm/cold instead of hot/hot water setting on clothes washer (electrically heated water) p. 82	5,040	0	0.0 _____
N: Add Desuperheater to geothermal system for domestic hot water, p. 113	4,949	1,150	5.8 _____
I: Add Desuperheater to geothermal system for domestic hot water (incremental), p. 113	4,949	695	3.5 _____
N: Buy electric-ignition tankless water heater, p. 115	4,751	728	5.9 _____
I: Choose electric-ignition tankless water heater when replacing broken unit, p. 115	4,751	123	1.0 _____
I: Replace electric dryer with gas dryer, p. 84	4,512	200	4.3 _____
N: Replace electric dryer with gas dryer (new), p. 84	4,512	600	13.0 _____
N: Replace 20 of your 25 lights with Compact Fluorescent Lights (CFLs), p. 56	4,489	117	1.7 _____
N: Put computer in sleep-mode, turn off when unused, p. 85	4,406	5	0.1 _____
N: Add manual chimney cap to fireplace flue, p. 112	4,138	510	4.8 _____
N: Add motorized chimney cap to fireplace flue, p. 112	4,138	1,160	10.8 _____
I: Replace electric stove with natural gas stove, p. 79	4,078	529	12.7 _____
N: Replace electric stove with natural gas stove (new), p. 79	4,078	150	3.6 _____
I: Replace electric dryer with gas dryer (washer is front-loader), p. 84	4,061	150	3.6 _____
I: Choose high-efficiency furnace instead of medium-efficiency when replacing, all other upgrades implemented, p.111	3,540	1,040	11.4 _____
N: Add R-40 cellulose insulation to unheated attic, p. 108	3,386	564	6.4 _____
N: Install window kits ("shrinkfoil) on 50% of your windows, p. 110	3,310	63	0.7 _____
I: Replace top-loading washer with front-loading model (water is heated by gas), p. 82	3,301	53	0.8 _____
N: Replace washer with front-loading model (water is heated by gas; new cost), p. 82	3,301	614	9.2 _____
N: Add R-20 insulation to basement walls, p. 108	3,009	1,403	18.0 _____
N: Get a tune-up for your furnace, p. 111	2,979	150	1.9 _____
N: Add R-12 insulation to basement walls, p. 108	2,837	1,005	13.7 _____
N: Add R-20 cellulose insulation to unheated attic, p. 108	2,565	292	4.4 _____
N: Add R-20 fibre-bat insulation to unheated attic, p. 108	2,565	620	9.4 _____
N: Use cold/cold instead of hot/hot wash mode (gas heated water) p. 82	2,267	0	0.0 _____
N: Add R-12 cellulose insulation to unheated attic, p. 108	1,924	193	5.5 _____
N: Add R-12 fibre-bat insulation to unheated attic, p. 108	1,924	373	7.5 _____
N: Use warm/cold instead of hot/hot water setting on clothes washer (gas heated water) p. 82	1,722	0	0.0 _____
N: Place R-11 insulating blanket on hot water tank, p. 116	1,451	15	0.4 _____
N: Install a solar tube in a frequently used area, p. 65	1,390	407	18.8 _____
N: Replace your Christmas lights with LEDs, p. 95	1,355	60	2.8 _____
N: Add motorized combustion air dampers to fresh-air intake for furnace, p. 111	1,324	426	12.5 _____
I: Self-booting, small off-grid, rural solar electricity (PV) installation, p. 103	1,129	-1,536	∞ _____
N: Purchase efficient cookware p. 80	1,058	259	9.5 _____
N: Buy two low-flush toilets to replace 7 gallon toilets, p. 121	940	300	5.1 _____
I: Purchase efficient cookware (incremental) p. 80	846	30	1.4 _____
N: Use cold/cold instead of hot/hot wash mode (gas heated front-loading washer) p. 82	756	0	0.0 _____
N: Buy two low-flush toilets to replace 3.5 gallon toilets, p. 121	662	450	10.9 _____
N: Use warm/cold instead of hot/hot wash mode (gas heated front-loading washer) p. 82	574	0	0.0 _____
N: Install toilet dams in your toilets to reduce flush volume, p. 121	218	27	2.0 _____
N: Insulate first 9 feet of water tank hot water pipe, p. 116	139	2	0.4 _____
N/I: Replace your old fridge (check page 75 to see if it is worthwhile for you)			_____

Your total Savings (lbs.) _____

Carbon Miser: 40.2%; $14,120.30

Appendix B: Highest CROI

Efficiency measures with the highest carbon return on investment (CROI) in pounds CO_2 saved per $ invested. I= incremental scenario (choose more efficient alternative when replacing worn-out model), N=new scenario (old unit still functional, replace with more efficient model).

Efficiency Measure	CO_2 CROI ratio (higher is better)	Life lbs. (higher is better)	Payback Years (lower is better)	Your Savings lbs. (higher is better)
I: Toyota Echo (manual) as primary family car (5 seats), p. 47	∞	112,125	0.0	
I: Toyota Echo (manual) as secondary family car (5 seats), p. 47	∞	42,482	0.0	
N: Use warm/cold instead of hot/hot water setting on clothes washer (gas heated water) p. 82	∞	8,608	0.0	
N: Use warm/cold instead of hot/hot water setting on clothes washer (electrically heated? water) p. 82	∞	25,198	0.0	
N: Use warm/cold instead of hot/hot wash mode (gas heated front-loading washer) p. 82	∞	2,869	0.0	
N: Use cold/cold instead of hot/hot wash mode (gas heated water) p. 82	∞	11,334	0.0	
N: Use cold/cold instead of hot/hot wash mode (electrically heated water) p. 82	∞	33,178	0.0	
N: Use cold/cold instead of hot/hot wash mode (gas heated front-loading washer) p. 82	∞	3,778	0.0	
I: Self-booting, small off-grid, rural solar electricity (PV) installation, p. 103	∞	11,285	∞	
N: Put computer in sleep-mode, turn off when unused, p. 85	4,406.4	22,032	0.1	
N: Subscribe to green power, p. 95	4,122.9	216,944	N/A	
N: Subscribe to green power (after you got geothermal heating/cooling/water heating), p. 95	4,122.9	239,800	N/A	
N: Subscribe to green power (geothermal, no fuel switching), p. 95	4,122.9	205,150	N/A	
N: Insulate first 9 feet of water tank hot water pipe, p. 116	371.5	557	0.4	
I: Replace top-loading washer with front-loading model (water is heated electrically), p. 82	293.5	15,448	0.4	
N: Place R-11 insulating blanket on hot water tank, p. 116	280.3	4,064	0.4	
N: Eliminate 90% of your power vampires, p. 67	269.3	33,664	1.2	
N: Replace your christmas lights with LEDs, p. 95	225.9	13,555	2.8	
N: Add R-12 cellulose insulation to unheated attic, p. 108	159.5	30,786	5.5	
I: Purchase efficient cookware (incremental) p. 80	152.4	4,571	1.4	
N: Add R-20 cellulose insulation to unheated attic, p. 108	140.6	41,047	4.4	
I: Replace top-loading washer with front-loading model (water is heated by gas), p. 82	138.0	7,263	0.8	
N: Replace electric stove with natural gas stove (new), p. 79	135.9	20,392	3.6	
N: Replace 20 of your 25 lights with Compact Fluorescent Lights (CFLs), p. 56	109.6	12,825	1.7	
I: Mercedes Smart Diesel car as secondary car (2 seats), p. 46	106.7	80,011	1.1	
I: Toyota Prius hybrid car as primary family car (5 seats), p. 42	105.0	149,649	1.1	
N: Add R-40 cellulose insulation to unheated attic, p. 108	96.1	54,183	6.4	
I: Choose electric-ignition tankless water heater when replacing broken unit, p. 115	85.0	10,453	1.0	
N: Add R-12 fibre-bat insulation to unheated attic, p. 108	82.5	30,786	7.5	
N: Add R-20 fibre-bat insulation to unheated attic, p. 108	66.2	41,047	9.4	
I: Off-grid, rural wind power installation, p. 102	60.2	71,301	3.0	
I: Replace electric dryer with gas dryer (washer is front-loader), p. 84	59.6	8,933	3.6	
N: Install a solar tube in a frequently used area, p. 65	54.6	22,248	18.8	
I: Honda Insight hybrid car (manual) as primary car (2 seats max.), p. 39	52.8	158,307	2.1	
I: Honda Insight hybrid car (auto) as secondary car (2 seats max.), p. 39	51.8	80,284	2.2	
I: Replace electric dryer with gas dryer, p. 84	49.6	9,926	4.3	
N: Seal air leaks in your house, p. 107	46.1	10,593	1.3	
N: Add R-12 insulation to basement walls, p. 108	45.2	45,398	13.7	
I: Replace electric stove with natural gas stove, p. 79	38.5	20,392	12.7	
N: Use 50% biodiesel for Smart or other diesel, p. 49	36.0	21,378	N/A	
I: Off-grid, rural solar electricity (PV) installation, p. 101	35.1	70,219	7.2	
N: Add R-20 insulation to basement walls, p. 108	34.3	48,149	18.0	

Carbon Buster: 72.6%; $17,805.00

N: Add manual chimney cap to fireplace flue, p. 112	32.5	16,551	4.8
N: Convert primary car to compressed natural gas, p. 48	30.5	61,030	3.0
I: Honda Insight hybrid car (manual) as secondary car (2 seats max.), p. 39	29.3	87,971	3.9
I: Add Desuperheater to geothermal system for domestic hot water (incremental), p. 114	28.5	19,795	3.5
I: Choose high-efficiency furnace instead of medium-efficiency when replacing, no other upgrades, p. 111	24.2	25,158	6.4
N: Purchase efficient cookware p. 80	22.1	5,713	9.5
N: Install window kits ("shrink foil") on 50% of your windows, p. 110	20.9	1,324	0.7
I: Off-grid, rural solar/wind hybrid power installation, p. 102	20.4	145,447	13.5
N: Get a tune-up for your furnace, p. 112	19.9	2,979	1.9
I: City-based, grid-connected solar electricity (PV) installation, p. 101	18.1	80,001	35.4
N: Add Desuperheater to geothermal system for domestic hot water, p. 114	17.2	19,795	5.8
I: Geothermal system for heating, AC, hot water, p. 113	16.6	97,551	6.8
N: Replace electric dryer with gas dryer (new), p. 84	16.5	9,926	13.0
N: Install toilet dams in your toilets to reduce flush volume, p. 121	16.2	437	2.0
I: Replace furnace and air conditioning with ground-source heat pump (geothermal), incremental, p. 113	16.0	77,756	7.4
N: Buy electric-ignition tankless water heater, p. 115	14.4	10,453	5.9
N: Add motorized chimney cap to fireplace flue, p. 112	14.3	16,551	10.8
I: Choose high-efficiency furnace instead of medium-efficiency when replacing, all other upgrades, p. 111	13.6	14,161	11.4
N: Buy two low-flush toilets to replace 7 gallon toilets, p. 121	12.5	3,761	5.1
N: Add motorized combustion air dampers to fresh-air intake for furnace, p. 111	12.4	5,296	12.5
N: Replace washer with front-loading model (water is heated by gas; new cost), p. 82	11.8	7,263	9.2
N: Geothermal system for heating, AC, hot water, p. 113	11.5	97,551	9.9
N: Install ground-source heat pump (geothermal), p. 113	10.4	77,756	11.4
N: Install a thermal solar collector for your hot water needs, p. 115	10.1	37,662	15.3
N: Geothermal system for heating, AC, hot water, after insulation, window kits etc., p. 113	7.5	63,792	19.2
N: Upgrade medium-efficiency to high-efficiency furnace, no other improvements, p. 111	6.9	25,158	22.6
N: Buy two low-flush toilets to replace 3.5 gallon toilets, p. 121	5.9	2,647	10.9
N: Upgrade medium-efficiency to high-efficiency furnace, all other improvements, p. 111	5.8	21,122	26.9
N/I: Replace your old fridge (check page 75 to see if it is worthwhile for you)			

Your total Savings (lbs.) _____

Appendix C: Best Investments

Efficiency measures with the highest internal rate of return (IRR; i.e. which is the best investment in monetary terms). I=incremental scenario (choose more efficient alternative when replacing worn-out model), N=new scenario (old unit still functional, replace with more efficient model). Negative numbers indicate increased consumption.

	Savings over life of measure		Payback	Your
	IRR %	Life $	Years	Savings $
	(higher	(higher	(lower	(higher
Efficiency Measure	is better)	is better)	is better)	is better)
I: Toyota Echo (manual) as primary family car (5 seats), p. 47	∞	15,990	0.0	
I: Toyota Echo (manual) as secondary family car (5 seats), p. 47	∞	6,058	0.0	
N: Use warm/cold instead of hot/hot water setting on clothes washer (gas heated water) p. 82	∞	1,112	0.0	
N: Use warm/cold instead of hot/hot water setting on clothes washer (electrically heated water) p. 82	∞	1,963	0.0	
N: Use warm/cold instead of hot/hot wash mode (gas heated front-loading washer) p. 82	∞	371	0.0	
N: Use cold/cold instead of hot/hot wash mode (gas heated water) p. 82	∞	1,464	0.0	
N: Use cold/cold instead of hot/hot wash mode (electrically heated water) p. 82	∞	2,584	0.0	

Carbon Miser: 40.2%; $14,120.30

N: Use cold/cold instead of hot/hot wash mode (gas heated front-loading washer) p. 82	∞	488	0.0	_____
I: Self-booting, small off-grid, rural solar electricity (PV) installation, p. 103	∞	8,883	∞	_____
N: Put computer in sleep-mode, turn off when unused, p. 85	1373%	1,716	0.1	_____
N: Place R-11 insulating blanket on hot water tank, p. 116	259%	525	0.4	_____
N: Insulate first 9 feet of water tank hot water pipe, p. 116	240%	72	0.4	_____
I: Replace top-loading washer with front-loading model (water is heated electrically), p. 82	229%	1,327	0.4	_____
I: Replace top-loading washer with front-loading model (water is heated by gas), p. 82	127%	734	0.8	_____
N: Install window kits ("shrink foil") on 50% of your windows, p. 110	102%	171	0.7	_____
I: Choose electric-ignition tankless water heater when replacing broken unit, p. 115	100%	1,350	1.0	_____
I: Toyota Prius hybrid car as primary family car (5 seats), p. 42	94%	21,341	1.1	_____
I: Mercedes Smart Diesel car as secondary car (2 seats), p. 46	93%	11,410	1.1	_____
N: Eliminate 90% of your power vampires, p. 67	84%	2,622	1.2	_____
N: Seal air leaks in your house, p. 107	73%	1,369	1.3	_____
I: Purchase efficient cookware (incremental) p. 80	73%	591	1.4	_____
N: Replace 20 of your 25 lights with Compact Fluorescent Lights (CFLs), p. 56	60%	999	1.7	_____
N: Install toilet dams in your toilets to reduce flush volume,p. 121	50%	136	2.0	_____
I: Honda Insight hybrid car (manual) as primary car (2 seats max.), p. 39	47%	22,576	2.1	_____
I: Honda Insight hybrid car (auto) as secondary car (2 seats max.), p. 39	45%	11,449	2.2	_____
N: Get a tune-up for your furnace, p. 112	43%	385	1.9	_____
N: Replace your Christmas lights with LEDs, p. 95	35%	1,056	2.8	_____
N: Convert primary car to compressed natural gas, p. 48	33%	11,077	3.0	_____
I: Off-grid, rural wind power installation, p. 102	33%	9,737	3.0	_____
N: Add Desuperheater to geothermal system for domestic hot water, p. 114	28%	3,980	5.8	_____
I: Add Desuperheater to geothermal system for domestic hot water (incremental), p. 114	28%	3,980	3.5	_____
N: Replace electric stove with natural gas stove (new), p. 79	28%	1,045	3.6	_____
I: Replace electric dryer with gas dryer (washer is front-loader), p. 84	25%	458	3.6	_____
I: Honda Insight hybrid car (manual) as secondary car (2 seats max.), p. 39	25%	12,546	3.9	_____
N: Add R-20 cellulose insulation to unheated attic, p. 108	23%	5,303	4.4	_____
N: Add manual chimney cap to fireplace flue, p. 112	20%	2,138	4.8	_____
I: Replace electric dryer with gas dryer, p. 84	20%	509	4.3	_____
N: Buy two low-flush toilets to replace 7 gallon toilets, p. 121	19%	1,174	5.1	_____
N: Add R-12 cellulose insulation to unheated attic, p. 108	18%	2,784	5.5	_____
N: Add R-40 cellulose insulation to unheated attic, p. 108	16%	7,000	6.4	_____
I: Choose high-efficiency furnace instead of medium-efficiency when replacing, no other upgrades implemented, p. 111	15%	3,250	6.4	_____
I: Off-grid, rural solar electricity (PV) installation, p. 101	14%	13,836	7.2	_____
I: Geothermal system for heating, AC, hot water, p. 113	13%	17,170	6.8	_____
N: Add R-12 fibre-bat insulation to unheated attic, p. 108	13%	3,977	7.5	_____
I: Replace furnace and air conditioning with ground-source heat pump (geothermal), incremental cost, p. 111	12%	13,190	7.4	_____
N: Buy electric-ignition tankless water heater, p. 116	12%	1,350	5.9	_____
N: Add R-20 fibre-bat insulation to unheated attic, p. 108	11%	5,303	9.4	_____
N: Purchase efficient cookware p. 80	10%	738	9.5	_____
N: Geothermal system for heating, AC, hot water, p. 113	8%	17,170	9.9	_____
N: Add R-12 insulation to basement walls, p. 108	7%	5,865	13.7	_____
I: Off-grid, rural solar/wind hybrid power installation, p. 102	7%	16,879	13.5	_____
N: Add motorized chimney cap to fireplace flue, p. 112	7%	2,138	10.8	_____
N: Buy two low-flush toilets to replace 3.5 gallon toilets, p. 121	7%	826	10.9	_____
I: Replace electric stove with natural gas stove, p. 79	6%	1,045	12.7	_____

Carbon Buster: 72.6%; $17,805.00

Efficiency Measure	%	$	IRR	Your Savings
I: Choose high-efficiency furnace instead of medium-efficiency when replacing, all other upgrades implemented, p. 111	6%	1,829	11.4	_____
N: Add R-20 insulation to basement walls, p. 108	5%	6,221	18.0	_____
N: Install a solar tube in a frequently used area, p. 65	5%	1,733	18.8	_____
N: Add motorized combustion air dampers to fresh-air intake for furnace, p. 112	5%	684	12.5	_____
N: Replace washer with front-loading model (water is heated by gas; new cost), p. 82	3%	734	9.2	_____
N: Install a thermal solar collector for your hot water needs, p. 115	3%	4,866	15.3	_____
I: City-based, grid-connected solar electricity (PV) installation, p. 101	1%	6,231	35.4	_____
N: Geothermal system for heating, AC, hot water, after insulation, window kits etc., p. 114	0.4%	8,848	19.2	_____
N: Use 50% biodiesel for Smart or other diesel, p. 50	0%	0	N/A	_____
N: Install ground-source heat pump (geothermal), p. 114	0%	13,190	11.4	_____
N: Upgrade medium-efficiency to high-efficiency furnace, no other improvements, p. 111	-1%	3,250	22.6	_____
N: Replace electric dryer with gas dryer (new), p. 84	-3%	509	13.0	_____
N: Upgrade medium-efficiency to high-efficiency furnace, all other improvements, p. 111	-3%	2,729	26.9	_____
N/I: Replace your old fridge (check pages 70 to 76 to see if this is worthwhile for you)				
N: Subscribe to green power, p. 96	N/A	0	N/A	_____
N: Subscribe to green power (after you got geothermal heating/cooling/water heating),p. 96	N/A	0	N/A	_____
N: Subscribe to green power (geothermal, no fuel switching), p. 96	N/A	0	N/A	_____

Your total Savings ($) _____

Appendix D: Highest Dollar Savings

Efficiency measures with highest dollar savings. I=incremental scenario (choose more efficient alternative when replacing worn-out model) , N=new scenario (old unit still functional, replace with more efficient model). Negative numbers indicate increased consumption.

Efficiency Measure	5 Years $ (higher is better)	IRR % (higher is better)	Your Savings
I: Honda Insight hybrid car (manual) as primary car (2 seats max.), p. 40	7,111	47%	_____
I: Toyota Prius hybrid car as primary family car (5 seats), p. 42	6,722	94%	_____
I: Toyota Echo (manual) as primary family car (5 seats), p. 47	5,037	∞	_____
N: Geothermal system for heating, AC, hot water, p. 114	4,292	8%	_____
I: Geothermal system for heating, AC, hot water, p. 114	4,292	13%	_____
I: Honda Insight hybrid car (manual) as secondary car (2 seats max.), p. 40	3,821	25%	_____
I: Honda Insight hybrid car (auto) as secondary car (2 seats max.), p. 40	3,487	45%	_____
I: Mercedes Smart Diesel car as secondary car (2 seats), p. 45	3,475	93%	_____
N: Convert primary car to compressed natural gas, p. 49	3,374	33%	_____
N: Install ground-source heat pump (geothermal), p. 114	3,297	0%	_____
I: Replace furnace and air conditioning with ground-source heat pump (geothermal), incremental cost, p. 114	3,297	12%	_____
I: Off-grid, rural solar/wind hybrid power installation, p. 103	2,640	7%	_____
N: Geothermal system for heating, AC, hot water, after insulation, window kits etc., p. 114	2,212	0.4%	_____
I: Off-grid, rural wind power installation, p. 102	1,947	33%	_____
I: Toyota Echo (manual) as secondary family car (5 seats), p. 47	1,845	∞	_____
I: Off-grid, rural solar electricity (PV) installation, p. 102	1,384	14%	_____
N: Install a thermal solar collector for your hot water needs, p. 115	1,216	3%	_____
N: Add Desuperheater to geothermal system for domestic hot water, p. 114	995	28%	_____
I: Add Desuperheater to geothermal system for domestic hot water (incremental), p. 114	995	28%	_____
I: Self-booting, small off-grid, rural solar electricity (PV) installation, p. 104	888	∞	_____
N: Seal air leaks in your house, p. 107	855	73%	_____
N: Upgrade medium-efficiency to high-efficiency furnace, no other improvements, p. 111	813	-1%	

Carbon Miser: 40.2%; $14,120.30

Action	$	%	
I: Choose high-efficiency furnace instead of medium-efficiency when replacing, no other upgrades implemented, p. 111	813	15%	_____
N: Upgrade medium-efficiency to high-efficiency furnace, all other improvements, p. 111	682	-3%	_____
I: City-based, grid-connected solar electricity (PV) installation, p. 101	623	1%	_____
N: Buy electric-ignition tankless water heater, p. 115	614	12%	_____
I: Choose electric-ignition tankless water heater when replacing broken unit, p. 115	614	100%	_____
I: Replace top-loading washer with front-loading model (water is heated electrically), p. 82	603	229%	_____
N: Install toilet dams in your toilets to reduce flush volume, p. 121	68	50%	_____
N: Add manual chimney cap to fireplace flue, p. 112	535	20%	_____
N: Add motorized chimney cap to fireplace flue, p. 112	535	7%	_____
N: Eliminate 90% of your power vampires, p. 67	524	84%	_____
N: Use cold/cold instead of hot/hot wash mode (electrically heated water) p. 82	517	∞	_____
I: Choose high-efficiency furnace instead of medium-efficiency when replacing, all other upgrades implemented, p. 111	457	6%	_____
N: Add R-40 cellulose insulation to unheated attic, p. 108	438	16%	_____
N: Install window kits ("shrink foil") on 50% of your windows, p. 110	428	102%	_____
N: Use warm/cold instead of hot/hot water setting on clothes washer (electrically heated water) p. 82	393	∞	_____
N: Add R-20 insulation to basement walls, p. 108	389	5%	_____
N: Get a tune-up for your furnace, p. 112	385	43%	_____
N: Add R-12 insulation to basement walls, p. 108	367	7%	_____
N: Replace 20 of your 25 lights with Compact Fluorescent Lights (CFLs), p. 56	350	60%	_____
N: Put computer in sleep-mode, turn off when unused, p. 85	343	1373%	_____
I: Replace top-loading washer with front-loading model (water is heated by gas), p. 82	334	127%	_____
N: Replace washer with front-loading model (water is heated by gas; new cost), p. 82	334	3%	_____
N: Add R-20 cellulose insulation to unheated attic, p. 108	331	23%	_____
N: Add R-20 fibre-bat insulation to unheated attic, p. 108	331	11%	_____
N: Buy two low-flush toilets to replace 7 gallon toilets, p. 121	294	19%	_____
N: Use cold/cold instead of hot/hot wash mode (gas heated water) p. 82	293	∞	_____
N: Add R-12 fibre-bat insulation to unheated attic, p. 108	249	13%	_____
I: Replace electric dryer with gas dryer, p. 84	231	20%	_____
N: Replace electric dryer with gas dryer (new), p. 84	231	-3%	_____
N: Use warm/cold instead of hot/hot water setting on clothes washer (gas heated water) p. 82	222	∞	_____
I: Replace electric stove with natural gas stove, p. 79	209	6%	_____
N: Replace electric stove with natural gas stove (new), p. 79	209	28%	_____
I: Replace electric dryer with gas dryer (washer is front-loader), p. 84	208	25%	_____
N: Buy two low-flush toilets to replace 3.5 gallon toilets, p. 121	207	7%	_____
N: Place R-11 insulating blanket on hot water tank, p. 116	188	259%	_____
N: Add R-12 cellulose insulation to unheated attic, p. 108	174	18%	_____
N: Add motorized combustion air dampers to fresh-air intake for furnace, p. 112	171	5%	_____
N: Purchase efficient cookware p. 80	137	10%	_____
I: Purchase efficient cookware (incremental) p. 80	109	73%	_____
N: Install a solar tube in a frequently used area, p. 65	108	5%	_____
N: Replace your Christmas lights with LEDs, p. 95	106	35%	_____
N: Use cold/cold instead of hot/hot wash mode (gas heated front-loading washer) p. 82	98	∞	_____
N: Use warm/cold instead of hot/hot wash mode (gas heated front-loading washer) p. 82	74	∞	_____
N: Insulate first 9 feet of water tank hot water pipe, p. 116	18	240%	_____
N: Use 50% biodiesel for Smart or other diesel, p. 49	0	0%	_____
N: Subscribe to green power, p. 95	0	N/A	_____
N: Subscribe to green power (after you get geothermal heating/cooling/water heating), p. 95	0	N/A	_____
N: Subscribe to green power (geothermal, no fuel switching), p. 95	0	N/A	_____
N/I: Replace your old fridge (check page 75 to see if it is worthwhile for you)			_____
	Your total Savings ($)		_____

Carbon Buster: 72.6%; $17,805.00

Appendix E: Complete Savings Details

Complete savings details. I=incremental scenario (choose more efficient alternative when replacing worn-out model), N=new scenario (old unit still functional, replace with more efficient model). Negative numbers indicate increased consumption, e.g., when switching from one fuel to another.

Efficiency Measure	Savings Carbon Pie %	CROI New	Payback Years	IRR %	Dollars Life $
Transportation					
I: Honda Insight hybrid car (manual) as secondary car (2 seats max.), p. 39	9.4%	29.3	3.9	25%	12,546
I: Honda Insight hybrid car (auto) as secondary car (2 seats max.), p. 39	8.6%	51.8	2.2	45%	11,449
I: Honda Insight hybrid car (manual) as primary car (2 seats max.), p. 39	17.5%	52.8	2.1	47%	22,576
I: Toyota Prius hybrid car as primary family car (5 seats), p. 42	16.5%	105.0	1.1	94%	21,341
I: Mercedes Smart Diesel car as secondary car (2 seats), p. 46	8.5%	106.7	1.1	93%	11,410
N: Use 50% biodiesel for Smart or other diesel, p. 49	7.5%	36.0	N/A	0%	0
I: Toyota Echo (manual) as primary family car (5 seats), p. 47	12.4%	∞	0.0	∞	15,990
I: Toyota Echo (manual) as secondary family car (5 seats), p. 47	4.5%	∞	0.0	∞	6,058
N: Convert primary car to compressed natural gas, p. 48	6.5%	30.5	3.0	33%	11,077
Electricity					
N: Replace 20 of your 25 lights with Compact Fluorescent Lights (CFLs), p. 56	1.6%	109.6	1.7	60%	999
N: Install a solar tube in a frequently used area, p. 65	0.5%	54.6	18.8	5%	1,733
N: Eliminate 90% of your power vampires, p. 67	2.4%	269.3	1.2	84%	2,622
N/I: Replace your old fridge (check page 75 to see if it is worthwhile for you)					
I: Replace electric stove with natural gas stove, p. 79	1.4%	38.5	12.7	6%	1,045
N: Replace electric stove with natural gas stove (new), p. 79	1.4%	135.9	3.6	28%	1,045
N: Purchase efficient cookware p. 80	0.4%	22.1	9.5	10%	738
I: Purchase efficient cookware (incremental) p. 80	0.3%	152.4	1.4	73%	591
I: Replace top-loading washer with front-loading model (water is heated by gas), p. 82	1.2%	138.0	0.8	127%	734
I: Replace top-loading washer with front-loading model (water is heated electrically), p. 82	2.5%	293.5	0.4	229%	1,327
N: Replace washer with front-loading model (water is heated by gas; new cost), p. 82	1.2%	11.8	9.2	3%	734
N: Use warm/cold instead of hot/hot water setting on clothes washer (gas heated water) p. 82	0.6%	∞	0.0	∞	1,112
N: Use warm/cold instead of hot/hot water setting on clothes washer (electrically heated water) p. 82	1.8%	∞	0.0	∞	1,963
N: Use warm/cold instead of hot/hot wash mode (gas heated front-loading washer) p. 82	0.2%	∞	0.0	∞	371
N: Use cold/cold instead of hot/hot wash mode (gas heated water) p. 82	0.8%	∞	0.0	∞	1,464
N: Use cold/cold instead of hot/hot wash mode (electrically heated water) p. 82	2.3%	∞	0.0	∞	2,584
N: Use cold/cold instead of hot/hot wash mode (gas heated front-loading washer) p. 82	0.3%	∞	0.0	∞	488
I: Replace electric dryer with gas dryer, p. 84	1.6%	49.6	4.3	20%	509
I: Replace electric dryer with gas dryer (washer is front-loader), p. 84	1.4%	59.6	3.6	25%	458
N: Replace electric dryer with gas dryer (new), p. 84	1.6%	16.5	13.0	-3%	509
N: Put computer in sleep-mode, turn off when unused, p. 85	1.5%	4,406.4	0.1	1373%	1,716
N: Replace your Christmas lights with LEDs, p. 95	0.5%	225.9	2.8	35%	1,056
N: Subscribe to green power, p. 95	15.2%	4,122.9	N/A	N/A	0
N: Subscribe to green power (after you get geothermal heating/cooling/ water heating), p. 95	16.8%	4,122.9	N/A	N/A	0

> 🎩 Carbon Miser: 40.2%; $14,120.30

5 Years $	CO₂ Life lbs.	CO₂ Life Tonnes	CO₂ 5 Years lbs.	CO₂ 5 Years Tonnes	Power kWh	Natural Gas 1kcu ft	Water gal	Gasoline gal	Cost New $	Incremental $	Your Savings $
3,821	87,971	39.9	26,793	12.2	0	0	0	274	19,330	3,000	____
3,487	80,284	36.4	24,452	11.1	0	0	0	250	20,080	1,550	____
7,111	158,307	71.8	49,867	22.6	0	0	0	510	19,330	3,000	____
6,722	149,649	67.9	47,139	21.4	0	0	0	462	20,275	1,425	____
3,475	80,011	36.3	24,369	11.1	0	0	0	249	750	750	____
0	21,378	9.7	21,378	9.7	0	0	0	0	594	594	____
5,037	112,125	50.9	35,319	16.0	0	0	0	361	10,985	0	____
1,845	42,482	19.3	12,939	5.9	0	0	0	132	10,986	0	____
3,374	61,030	27.7	18,588	8.4	0	-89	0	741	2,000	2,000	____
350	12,825	5.8	4,489	2.0	707	0	0	0	135	117	____
108	22,248	10.1	1,390	0.6	219	0	0	0	410	407	____
524	33,664	15.3	6,733	3.1	1,060	0	0	0	125	125	____
209	20,392	9.2	4,078	1.8	976	-4	0	0	529	529	____
209	20,392	9.2	4,078	1.8	976	-4	0	0	150	150	____
137	5,713	2.6	1,058	0.5	0	2	0	0	259	259	____
109	4,571	2.1	846	0.4	0	1	0	0	30	30	____
334	7,263	3.3	3,301	1.5	108	3	7,410	0	53	53	____
603	15,448	7.0	7,022	3.2	1,068	0	7,410	0	53	53	____
334	7,263	3.3	3,301	1.5	108	3	7,410	0	614	614	____
222	8,608	3.9	1,722	0.8	0	3	0	0	0	0	____
393	25,198	11.4	5,040	2.3	794	0	0	0	0	0	____
74	2,869	1.3	574	0.3	0	1	0	0	0	0	____
293	11,334	5.1	2,267	1.0	0	4	0	0	0	0	____
517	33,178	15.0	6,636	3.0	1,045	0	0	0	0	0	____
98	3,778	1.7	756	0.3	0	1	0	0	0	0	____
231	9,926	4.5	4,512	2.0	1,079	-4	0	0	200	200	____
208	8,933	4.1	4,061	1.8	971	-3	0	0	150	150	____
231	9,926	4.5	4,512	2.0	1,079	-4	0	0	600	600	____
343	22,032	10.0	4,406	2.0	694	0	0	0	5	5	____
106	13,555	6.1	1,355	0.6	213	0	0	0	60	60	____
0	216,944	98.4	43,389	19.7		0	0	0	53	53	____
0	239,800	108.8	47,960	21.8		0	0	0	58	58	____

Carbon Buster: 72.6%; $17,805.00

Efficiency Measure	Savings Carbon Pie %	CROI New	Payback Years	IRR %	Dollars Life $
N: Subscribe to green power (geothermal, no fuel switching), p. 95	14.4%	4,122.9	N/A	N/A	0
I: Off-grid, rural solar electricity (PV) installation, p. 101	2.5%	35.1	7.2	14%	13,836
I: City-based, grid-connected solar electricity (PV) installation, p. 101	2.8%	18.1	35.4	1%	6,231
I: Off-grid, rural wind power installation, p. 102	5.0%	60.2	3.0	33%	9,737
I: Off-grid, rural solar/wind hybrid power installation, p. 102	8.0%	20.4	13.5	7%	16,879
I: Self-booting, small off-grid, rural solar electricity (PV) installation, p. 103	0.4%	∞	∞	∞	8,883
Heating					
N: Seal air leaks in your house, p. 107	2.3%	46.1	1.3	73%	1,369
N: Add R-40 cellulose insulation to unheated attic, p. 108	1.2%	96.1	6.4	16%	7,000
N: Add R-20 cellulose insulation to unheated attic, p. 108	0.9%	140.6	4.4	23%	5,303
N: Add R-20 fibre-bat insulation to unheated attic, p. 108	0.9%	66.2	9.4	11%	5,303
N: Add R-12 cellulose insulation to unheated attic, p. 108	0.7%	159.5	5.5	18%	2,784
N: Add R-12 fibre-bat insulation to unheated attic, p. 108	0.7%	82.5	7.5	13%	3,977
N: Add R-20 insulation to basement walls, p. 108	1.1%	34.3	18.0	5%	6,221
N: Add R-12 insulation to basement walls, p. 108	1.0%	45.2	13.7	7%	5,865
N: Install window kits ("shrink foil") on 50% of your windows, p. 110	1.2%	20.9	0.7	102%	171
N: Upgrade medium-efficiency to high-efficiency furnace, all other improvements, p. 111	1.9%	5.8	26.9	-3%	2,729
N: Upgrade medium-efficiency to high-efficiency furnace, no other improvements, p. 111	2.2%	6.9	22.6	-1%	3,250
I: Choose high-efficiency furnace instead of medium-efficiency when replacing, all other upgrades implemented, p. 111	1.2%	13.6	11.4	6%	1,829
I: Choose high-efficiency furnace instead of medium-efficiency when replacing, no other upgrades implemented, p. 111	2.2%	24.22	6.4	15%	3,250
N: Get a tune-up for your furnace, p. 112	1.0%	19.9	1.9	43%	385
N: Add motorized combustion air dampers to fresh-air intake for furnace, p. 112	0.5%	12.4	12.5	5%	684
N: Add manual chimney cap to fireplace flue, p. 112	1.5%	32.5	4.8	20%	2,138
N: Add motorized chimney cap to fireplace flue, p. 112	1.5%	14.3	10.8	7%	2,138
N: Install ground-source heat pump (geothermal), p. 113	6.8%	10.4	11.4	0%	13,190
I: Replace furnace and air conditioning with ground-source heat pump (geothermal), incremental cost, p. 113	6.8%	16.0	7.4	12%	13,190
N: Add Desuperheater to geothermal system for domestic hot water, p. 114	1.7%	17.2	5.8	28%	3,980
I: Add Desuperheater to geothermal system for domestic hot water (incremental), p. 114	1.7%	28.5	3.5	28%	3,980
N: Geothermal system for heating, AC, hot water, p. 113	8.6%	11.5	9.9	8%	17,170
I: Geothermal system for heating, AC, hot water, p. 113	8.6%	16.6	6.8	13%	17,170
N: Geothermal system for heating, AC, hot water, after insulation, window kits etc., p. 113	5.6%	7.5	19.2	0.4%	8,848
N: Install a thermal solar collector for your hot water needs, p. 116	3.3%	10.1	15.3	3%	4,866
N: Buy electric-ignition tankless water heater, p. 115	1.7%	14.4	5.9	12%	1,350
I: Choose electric-ignition tankless water heater when replacing broken unit, p. 115	1.7%	85.0	1.0	100%	1,350
N: Place R-11 insulating blanket on hot water tank, p. 116	0.5%	280.3	0.4	259%	525
N: Insulate first 9 feet of water tank hot water pipe, p. 116	0.0%	371.5	0.4	240%	72
Water					
N: Buy two low-flush toilets to replace 7 gallon toilets, p. 121	0.3%	12.5	5.1	19%	1,174
N: Buy two low-flush toilets to replace 3.5 gallon toilets, p. 121	0.2%	5.9	10.9	7%	826
N: Install toilet dams in your toilets to reduce flush volume, p. 121	0.1%	16.2	2.0	50%	136

Carbon Miser: 40.2%; $14,120.30

5 Years $	CO2 Life lbs.	CO2 Life Tonnes	CO2 5 Years lbs.	CO2 5 Years Tonnes	Power kWh	Natural Gas 1kcu ft	Water gal	Gasoline gal	Cost New $	Incremental $	Your Savings $
0	205,150	93.1	41,030	18.6		0	0	0	50	50	___
1,384	70,219	31.9	7,022	3.2	1,260	0	0	-10	7,999	1,999	___
623	80,001	36.3	8,000	3.6	1,260	0	0	0	4,415	4,415	___
1,947	71,301	32.3	14,260	6.5	2,400	0	0	-10	7,185	1,185	___
2,640	145,447	66.0	22,749	10.3	3,660	0	0	-5	13,119	7,119	___
888	11,285	5.1	1,129	0.5	420	0	0	-16	4,464	-1,536	___
855	10,593	4.8	6,621	3.0	0	11	0	0	230	230	___
438	54,183	24.6	3,386	1.5	0	6	0	0	564	564	___
331	41,047	18.6	2,565	1.2	0	4	0	0	292	292	___
331	41,047	18.6	2,565	1.2	0	4	0	0	620	620	___
174	30,786	14.0	1,924	0.9	0	3	0	0	193	193	___
249	30,786	14.0	1,924	0.9	0	3	0	0	373	373	___
389	48,149	21.8	3,009	1.4	0	5	0	0	1,403	1,403	___
367	45,398	20.6	2,837	1.3	0	5	0	0	1,005	1,005	___
428	1,324	0.6	3,310	1.5	0	5	0	0	63	63	___
682	21,122	9.6	5,281	2.4	0	9	0	0	3,667	3,667	___
813	25,158	11.4	6,290	2.9	0	10	0	0	3,667	3,667	___
457	14,161	6.4	3,540	1.6	0	6	0	0	1,040	1,040	___
813	25,158	11.4	6,290	2.9	0	10	0	0	1,040	1,040	___
385	2,979	1.4	2,979	1.4	0	5	0	0	150	150	___
171	5,296	2.4	1,324	0.6	0	2	0	0	426	426	___
535	16,551	7.5	4,138	1.9	0	7	0	0	510	510	___
535	16,551	7.5	4,138	1.9	0	7	0	0	1,160	1,160	___
3,297	77,756	35.3	19,439	8.8	-2,413	58	0	0	7,500	7,500	___
3,297	77,756	35.3	19,439	8.8	-2,413	58	0	0	4,873	4,873	___
995	19,795	9.0	4,949	2.2	-1,091	20	0	0	1,150	1,150	___
995	19,795	9.0	4,949	2.2	-1,091	20	0	0	1,150	695	___
4,292	97,551	44.2	24,388	11.1	-3,504	77	0	0	0	8,500	___
4,292	97,551	44.2	24,388	11.1	-3,504	77	0	0	0	5,873	___
2,212	63,792	28.9	15,948	7.2	-466	31	0	0	0	8,500	___
1,216	37,662	17.1	9,415	4.3	0	16	0	0	3,715	3,715	___
614	10,453	4.7	4,751	2.2	0	8	0	0	728	728	___
614	10,453	4.7	4,751	2.2	0	8	0	0	578	123	___
188	4,064	1.8	1,451	0.7	0	2	0	0	15	15	___
18	557	0.3	139	0.1	0	0	0	0	2	2	___
294	3,761	1.7	940	0.4	0	0	28,890	0	300	300	___
207	2,647	1.2	662	0.3	0	0	20,330	0	450	450	___
68	437	0.2	218	0.1	0	0	6,710	0	156	156	___

Total Savings ($)   ___

Carbon Buster: 72.6%; $17,805.00

Appendix F: Typical Household Energy Consumption

Energy consumption of a typical American family with four family members, two cars, and a 2,000 to 2,500 sq. ft. free-standing house (excludes indirect/upstream energy consumption, emissions, and costs except for electricity production).[109]

		lbs. CO2	$	
Total		57,047	6,255	

Cars		lbs. CO2	Gallons	$	Percent of category
	Car total	26,920	1,376	3,839	100
	SUV	17,255	882	2,461	64
	Sedan	9,665	494	1,378	36

Electricity		lbs. CO2	kWh	$	Percent of category
	Electricity total	19,323	11782	1019	100
	Air conditioning	4,566	2,784	241	24
	Rest	2,499	1,524	132	13
	Fridge	2,032	1,239	107	11
	Lights	1,932	1,178	102	10
	Power vampires	1,932	1,178	102	10
	Dryer	1,770	1,079	93	9
	Stove	1,600	976	84	8
	Freezer	840	512	44	4
	TV, large-screen	525	320	28	3
	Computer	430	262	23	2
	Microwave	343	209	18	2
	Washer	197	120	10	1.0
	TV, regular	196	120	10	1.0
	Hi-Fi, Video, DVD	115	70	6	0.6
	Ceiling Fan	82	50	4	0.4
	Alarm Clocks	74	45	4	0.4
	Printer	74	45	4	0.4
	Answering machine	57	35	3	0.3
	Portable phone	43	26	2	0.2
	Cell phone charger	15	9	1	0.1

Natural gas		lbs. CO2	cu. ft.	$	Percent of category
	Natural gas total	9,027	74,600	1,162	100
	Space heating	6,643	54,900	855	74
	Water heating	2,384	19,700	307	26

Air Travel		lbs. CO2	gal.	$	Percent of category
	Air travel total	1,080	–	–	100

Water		lbs. CO2	gal	$	Percent of category
	Water total	697	106,990	219	100

Carbon Miser: 40.2%; $14,120.30

Appendix G: Average Energy Prices

Average US national energy prices (at time of writing 2005/2006).[110]

Electricity (kWh)	¢9.89
Natural gas (1000 cu. ft.)	$15.58
Gasoline (gal)	$2.79
Premium unleaded gasoline (gal)	$3.00
Diesel Fuel (#2, gal)	$2.50
Biodiesel	$2.99
Propane (gal)	¢93.2
Water (gal)	¢0.203

Appendix H: CO_2 Emission Factors

Average national carbon dioxide emissions per unit of energy consumed (US averages unless otherwise indicated. CDN= Canadian. Direct emissions only; includes power plant emissions in case of electricity).[111]

Energy	CO_2
Natural Gas (million Btu)	117.08 lbs
Natural gas (1,000 cu. ft.)	120.59 lbs
Electricity (kWh; US)	1.27 lbs
Electricity (kWh; US)	0.576 kg
Electricity (kWh; CDN)	0.208 lbs
Diesel* (million Btu)	161.39 lbs
Diesel (US gallon)	22.38 lbs
Diesel (L)	5.913 lbs
Diesel (L)	2.68 kg
Kerosene (US gallon)	21.54 lbs
LPG (liquefied petroleum gas; US gallon)	12.805 lbs
Water (m3)	0.78 kg
Water (m3)	0.454 lbs
Water (gal)	0.00651 lbs
Biodiesel (US gallon)	4.924 lbs
Gasoline (US gallon)	19.564 lbs
Gasoline (US gallon)	8.874 kg
Gasoline (L)	2.34 kg
Propane (US gallon)	12.669 lbs
Propane (million Btu)	139.178 lbs

Note: *based on fuel oil.

Carbon Buster: 72.6%; $17,805.00

Appendix I: CO2 Emissions from Electricity Production

Average US State and Canadian provincial carbon dioxide emissions per kWh of electricity consumed (includes power plant emissions, excludes upstream emissions; CO_2e = carbon dioxide equivalent).[112]

United States

State	CO2 lbs./kWh	Ranking	State	CO2 lbs./kWh	Ranking
Alabama	1.31	24	Montana	1.43	30
Alaska	1.38	27	Nebraska	1.40	29
Arizona	1.05	13	Nevada	1.52	33
Arkansas	1.29	21	New Hampshire	0.68	6
California	0.61	5	New Jersey	0.71	7
Colorado	1.93	43	New Mexico	2.02	47
Connecticut	0.94	12	New York	0.86	11
Delaware	1.83	40	North Carolina	1.24	18
Florida	1.39	28	North Dakota	2.24	50
Georgia	1.37	25	Ohio	1.80	39
Hawaii	1.66	36	Oklahoma	1.72	38
Idaho	0.03	1	Oregon	0.28	4
Illinois	1.16	15	Pennsylvania	1.26	19
Indiana	2.08	48	Rhode Island	1.05	14
Iowa	1.88	42	South Carolina	0.83	9
Kansas	1.68	37	South Dakota	0.80	8
Kentucky	2.01	46	Tennessee	1.30	23
Louisiana	1.18	17	Texas	1.46	31
Maine	0.85	10	Utah	1.93	44
Maryland/D.C.	1.37	26	Vermont	0.03	2
Massachusetts	1.28	20	Virginia	1.16	16
Michigan	1.58	34	Washington	0.25	3
Minnesota	1.52	32	West Virginia	1.98	45
Mississippi	1.29	22	Wisconsin	1.64	35
Missouri	1.84	41	Wyoming	2.15	49

US average 1.34
For the years 1998–2000

Canada

Province/Territory	CO2e kg/kWh	Ranking
Alberta	0.985	13
British Columbia	0.032	3
Manitoba	0.018	2
New Brunswick	0.571	8
Newfoundland & Labrador	0.286	6
NWT/Nunavut	0.2	5
Nova Scotia	0.916	12
Ontario	0.304	7
Prince Edward Island	0.72	9
Quebec	0.0093	1
Saskatchewan	0.888	11
Yukon		
- diesel	0.765	10
- system	0.044	4

Canadian average 0.219
For the year 2002

Carbon Miser: 40.2%; $14,120.30

Appendix J: Energy Conversion Factors [113]

Operation; do this to the unit on the left using the conversion factor to get the unit on the right.

From	Conversion factor	Operation	To
Btu (mean)	1,055.87	multiply	J (joule)
kWh	3,600,000	multiply	J (joule)
kWh	3,409.51	multiply	Btu
Btu	3,409.51	divide	kWh
mpg	235.215	(divide 235.215 by the other)	L/100 km
L/100 km	235.215	(divide 235.215 by the other)	mpg
mpig (imperial)	278	multiply	L/100 km
L/100 km	278	multiply	mpg
kWh	3.6	multiply	MJ (million Joule)
MJ	3.6	divide	kWh
calorie	4.19	multiply	joule (mean)
GJ	277.7	multiply	kWh
therm (US)	0.1055	multiply	GJ
therm (US)	29.29	multiply	kWh
GJ Natural gas	1.05	divide	1k cu. ft.
1k cu. ft. Natural gas	278	multiply	kWh
Gasoline (gallon)	36.68	multiply	kWh
Gasoline (gallon)	125,071	multiply	Btu
kWh/m^2	316.75	multiply	Btu/sq. ft.
gasoline	0.121	multiply	natural gas 1k cu. ft.
1k cu. ft. Natural gas	1.05	multiply	GJ
Propane (US gallon)	26.7	multiply	kWh
Propane (US gallon)	10.986	divide million	Btu
Propane (million Btu)	293.3	multiply	kWh

Appendix K: Metric Conversions [114]

	From	Conversion factor	Operation	To
Weights	pound (avoirdupois)	0.4535924	multiply	kg
	kg	0.4535924	divide	lbs
	pound	0.000453592	multiply	metric tonne
Volumes	gal (US)	3.785412	multiply	L
	gal (Imp.)	4.54609	multiply	L
	cu. ft.	0.00003	multiply	L
	gal (US)	264.1720373	multiply	m^3
	m^3	264.1720373	multiply	US gal
Distances	miles	1.609344	multiply	km
	inches	2.54	multiply	cm
Areas	square feet	10.76391042	divide	m^2
	m^2	10.76391042	multiply	square feet
	square feet	9	divide	square yards

Carbon Buster: 72.6%; $17,805.00

Appendix L: How to Correct for Local Energy Prices

To get savings figures that reflect your local (and current) energy prices, take your personal energy rate for the affected utility, divide it by the national average cost used in this book (see Appendix G), and multiply it by the predicted savings provided in the recommendations of this book, or in the Green Checklists (Appendices A to E). Voilà! Your own personalized savings predictions!

In calculating long-term paybacks, keep in mind that energy prices are widely expected to continue to increase.

$$\text{Your savings} = \frac{\text{predicted savings from the book} * \text{your energy price}}{\text{average energy price used in book}}$$

Example:

Predicted energy savings:...................... $100
Electricity rate used in book:................ 9.89¢/kWh
Your electricity rate:............................ 12.2¢/kWh
Your savings... = $100*12.2¢/kWh/9.89¢/kWh=$123

Appendix M: For More Information

Suggested Books and Magazines for Further Reading

Books

ENERGY EFFICIENCY

- American Council for an Energy-Efficient Economy (ACEEE)/Alex Wilson, Jennifer Thorne, and John Morrill. *Consumer Guide to Home Energy Savings. 8th Ed.* ACEEE, 2003. $ 8.95.
  The gold standard of guides to the most energy efficient home appliances, plus other information on home energy. Alex Wilson is also editor of *Environmental Building News* (buildinggreen.com), a monthly newsletter for green building professionals.
- Paul Scheckel. *The Home Energy Diet: How to Save Money by Making your House Energy-Smart.* New Society Publishers, 2005. $18.95.
  Nuts and bolts of energy efficient living, from the perspective of an experienced home-energy auditor.
- William H. Kemp. S*mart Power: An Urban Guide to Renewable Energy and Efficiency.* Aztext Press, 2004. $29.95.
  A good general guide to efficiency, with special emphasis on solar and wind.
- Richard Heede and RMI Staff. *Homemade Money: How to Save Energy and Dollars in your Home.* Rocky Mountain Institute, 1995. $3-9, used.
  A practical guide to home energy savings. Getting a little dated, but still lots of good data, including information on draft-proofing your house.

THERMAL MASS BUILDING TECHNIQUES USING COB

- Ianto Evans, Michael G. Smith and Linda Smiley. *The Hand-Sculpted House: A Philosophical and Practical Guide to Building a Cob Cottage.* Chelsea Green Publishing, 2002. $35.00.
  A complete and well-illustrated natural building primer.

 Carbon Miser: 40.2%; $14,120.30

BACKGROUND INFORMATION ON THE ENERGY EFFICIENCY REVOLUTION

- Paul Hawken, Amory B. Lovins and L. Hunter Lovins. *Natural Capitalism: The Next Industrial Revolution.* Earthscan Publications, 2000. $17.95.
- Amory B. Lovins, E. Kyle Datta, Odd-Even Bustnes, Jonathan G. Koomey and Nathan J. Glasgow. *Winning the Oil Endgame: Innovation for Profit, Jobs and Security.* Earthscan Publications, 2005. (Available for free download from: oilendgame.com/ReadTheBook.html.

WIND

Paul Gipe. *Wind Power: Renewable Energy for Home, Farm and Business.* Chelsea Green Publishing, 2004. $50.00.

SOLAR WATER HEATING

Bob Ramlow with Benjamin Nusz. *Solar Water Heating: A Comprehensive Guide to Solar Water and Space Heating Systems.* New Society Publishers, 2006. $24.95

Magazines

- BackHome Magazine. backhomemagazine.com. $21.97/year (six issues). Good how-to information on green and sustainable building, renewable energy, and organic gardening.
- Home Power Magazine. homepower.com. $ 22.50/year (six issues). This is the one our office subscribes to.
- Home Energy Magazine. homeenergy.org. $65.00/year (six issues). Also good, but three times more expensive than Home Power Magazine.
- Mother Earth News. motherearthnews.com. $19.95/year (eight issues). Consumer information on natural building, renewable energy and sustainable agriculture.

Suggested Web Sites

- For updates to this book: carbonbusters.org/handbook.
- Rocky Mountain Institute Home Energy Briefs. rmi.org/sitepages/ pid171.php#LibHshldEnEff. Top-notch efficiency advice — for free!
- Energy Solutions: Office of Energy Efficiency and Renewable Energy, US Department of Energy. eere.energy.gov/buildings/info/homes/index.html. Energy solutions for your home.
- Energy Savers: Office of Energy Efficiency and Renewable Energy, US Department of Energy. eere.energy.gov/consumer/tips Information and tips for saving energy at home.
- Office of Energy Efficiency, Natural Resources Canada. oee.nrcan.gc.ca/energuide/ home.cfm. Canada's home energy rating system for appliances and houses. Note: all Canadian Government publications and web sites are also available in French.
- Energy Star Program. energystar.com. Detailed information on the most efficient devices rated by the Energy Star Partnership Program.
- Alliance to Save Energy (with sponsors). energyhog.org. A web site for kids only, plus information for adults on home energy conservation.

Carbon Buster: 72.6%; $17,805.00

Glossary of Terms

Albedo: The ratio of incoming visible light to reflected visible light from a surface. High albedo means more light is reflected (e.g.: a white surface has a high albedo).

Biodiesel: A fuel produced from oil-rich plant seeds that can be used instead of fossil-fuel-derived diesel.

Biofuels: Non-historic fuels produced from a biological source.

Carbon: One of the elements, found in all organisms. All fossil fuels contain carbon, and their burning releases the major greenhouse gas (see greenhouse gases) carbon dioxide.

CFL: Compact fluorescent light. A light with the efficiency gains of fluorescent tubes compressed into the approximate size of a conventional light bulb.

Composting toilet: A device that converts human waste into useful garden soil through a biological process of decomposition.

Coolth: Coolness, the state of being cool.

CROI: Carbon Return on Investment. The amount of greenhouse gas emission prevented (in kilograms, or other weight unit of carbon dioxide equivalents, or CO_2e) per monetary unit (e.g.: dollar or Euro).

Ecological footprint: The amount of area required to sustain the ecological processes of an entity, e.g.: a person or city. The ecological footprint includes the area required to grow food, or to metabolize carbon dioxide emissions. The term was coined by Canadian ecologists William Rees and Mathias Wackernagel in 1996.

Emission: The discharge of a polluting substance, most commonly applied to gases.

Emissions, direct: Emissions arising through the combustion of a fuel. Examples are sulfur dioxide and carbon dioxide.

Emissions, indirect: Emissions that arise not at the site of energy use, but geographically removed from it.

Emissions, particulate: Emissions of solid particles, e.g.: tiny soot particles from diesel engines.

Emissions, upstream: Emissions that arise in the production of a fuel, prior to its consumption.

Energy, primary: Energy used directly through the consumption of a fuel, either to provide an energy service directly, or to produce more convenient or cleaner secondary energy. Examples of primary energy use are the combustion of natural gas to heat a house, or the burning of coal for a power plant steam turbine to produce electricity.

Energy, secondary: A cleaner or more convenient form of energy created through the consumption of primary energy.

EROI: Energy Return on Investment. The amount of energy recovered divided by the amount of energy invested to obtain the energy. An EROI of better than one indicates that the process yields energy, an EROI of less than one indicates that the energy recovery process uses up more energy than it yields. For example, according to a 1984 Science study, the EROI of oil and gas discoveries was over 100 in the 1940s, but only eight in the 1970s.

Fixed costs: The component of your utility bills unaffected by your utility consumption. Fixed costs may include administration fees, meter reading fees, peak demand fees and wire service provider or other distribution fees.

Fuel cell: A device that produces electricity and heat by combining hydrogen and oxygen in a controlled reaction across a membrane. In contrast to combustion of fossil fuels, fuel cells produce no harmful emissions.

Full-spectrum lighting: A light that emits all or most of the components of natural sunlight.

Geothermal: The use of ground source heat or coolth for heating or cooling purposes. Geothermal power is the industrial use of naturally occurring hot water at shallow depths, restricted to favorable geographic locations. Geothermal heating, found in many American homes, is the use of the relatively stable temperature of the earth at a depth of 6 feet or more for cooling and heating purposes, using a ground source heat pump. Geothermal heating and cooling can be used at almost any site in the US (except parts of Alaska) and southern Canada.

Greenhouse gases: Any of the gases contributing to the natural or anthropogenic greenhouse effect, increasing the earth's temperature by reflecting infrared radiation back to the planet. The principal greenhouse gases include carbon dioxide, methane, water vapor, nitrous oxides, and chlorinated fluorocarbons.

Grey water: Household sewage exclusive of human waste, e.g., water from the kitchen sink, shower or washing machine.

HEV: Hybrid Electric Vehicle. A car that uses an electric as well as internal combustion engine for propulsion, to maximize efficiency.

Hypercar: A car that achieves energy efficiency gains by orders of magnitude compared to conventional cars through radical redesign. Methods employed include super-light construction materials, such as carbon-fiber reinforced polymer (C-FRP), hybrid gasoline-electric engines, reduced drag coefficient and regenerative braking.

Intertie: A connection between a (usually renewable) micro-power generation source and a utility electrical grid.

Inverter: A device that converts electricity from one voltage to another. In an off-grid residential setting, typically a device that converts 12-volt electricity to 120-volt electricity.

IRR: Internal Rate of Return, equivalent to the interest rate earned over the lifetime of an investment. It is calculated iteratively by taking the lifetime savings of a measure, subtracting the cost of implementation, and calculating the percent return per year of the remaining money over the life of the project.

LED: Light Emitting Diode. An efficient technology for producing point- sources of light.

Life-cycle costs: In the context of this book, the cradle-to-grave environmental costs that arise over the full use of a product, from the mining and harvesting operations to obtain the raw materials for production, over manufacturing, distribution, consumption and disposal.

Naturescaping: Landscaping making use of native vegetation, especially with the purpose of providing a variety of habitats for native plants and animals.

Payback: The time in years it takes for an energy efficiency investment to pay for itself out of energy savings. For example, a lamp that costs $100 to install and saves $25 per year in reduced energy costs has a payback of four years. Simple payback does not take into consideration the cost of (borrowing) money or inflation.

PHEV: Plug-in Hybrid Electric Vehicle. A hybrid electric vehicle (see HEV) that extends its range by plugging into an electrical outlet. Typically, it has a larger battery system for energy storage than an HEV.

PV: Photovoltaics, or electricity produced by solar-electric cells or modules.

R-value: A measure of thermal resistance, or thermal insulance, symbol "R". The higher the R-value, the better the insulating value. Commonly used to describe the insulating value of walls in houses. Inverse of U-value.

Imperial Units:

$$R = \frac{ft^2 \times °F \times h}{Btu}$$

Metric Units:

$$R = \frac{m^2 \times °C}{W}$$

Shrinkfoil: A foil whose surface area gets diminshed through the application of heat. Used to increase the insulating value of windows in home energy conservation.

Sustainable: Capable of being continued indefinitely. A process is considered sustainable when it does not endanger the capability of future generations to employ the same process.

Tonne: Metric tonne or ton, 1,000 kilograms. Equivalent to 2,204.62 avoirdupois (American) pounds. Symbol "t".

U-value: A measure of thermal transmittance, symbol "U". The lower the transmittance, the better the insulating value. Commonly used to describe the insulating value of windows. Inverse of R-value, i.e., $U = 1/R$.

Watt-hour: The number of watts consumed over one hour.

Xeriscaping: Landscaping using drought-resistant plants and materials that require minimal amounts of water for maintenance.

Endnotes

Congratulations!

1   Energy Information Administration. July 25, 2005. "International Total Primary Energy Consumption and Energy Intensity: All Countries, 1980-2003," table1g.xls. Retrieved March 13, 2006 from eia.doe.gov/emeu/international/energyconsumption.html. The primary energy consumption per US$2,000 gross domestic product (GDP) is 9,521 Btu for the US, vs. 6,427 Btu for the UK and 4,605 Btu for Japan in the year 2003, the latest year for which data are currently available.

2   Economic History Services. Johnston, Louis D. and Samuel H. Williamson. October 2005. "The Annual Real and Nominal GDP for these United States, 1790 - Present." Retrieved March 29, 2006 from
eh.net/hmit/gdp/gdp_answer.php?CHKnominalGDP=on&year1=1998&year2=2004.

3   Energy Information Administration. April 10, 2001. "Energy Price Impacts on the US Economy." Retrieved March 29, 2006, from eia.doc.gov/olaf/economy/energy-price.html.

Chapter 1: Carbon Busting for Fun and Profit

4   BBC News. Feb. 16, 2005. Q&A: "The Kyoto Protocol." Retrieved July 13, 2005 from news.bbc.co.uk/1/hi/sci/tech/4269921.stm.

5   United Nations Framework Convention on Climate Change. 2006. "Kyoto Protocol to the United Nations Framework Convention on Climate Change." Retrieved March 13, 2006 from unfccc.int/resource/docs/convkp/kpeng.pdf.

6   Global Climate Coalition. 2005. "Climate Economics: EIA Study Says Higher Energy Prices will be Result of U.S. Meeting Kyoto Targets." Retrieved July 17, 2005 from globalclimate.org.

7   Energy Information Administration. April 2005. "Impacts of Modeled Recommendations of the National Commission on Energy Policy." Retrieved March 13, 2006 from eia.doe.gov/oiaf/servicerpt/bingaman/pdf/sroiaf(2005)02.pdf.

8   San Diego Source. John Heilprin, Associated Press. April 15, 2005. "Energy Study Finds Greenhouse Gas Limits Affordable." Retrieved March 13, 2006 from sddt.com/News/article.cfm?SourceCode=2005041510.

9   Amory B. Lovins, and H. L. Lovins, 1991. *Least-Cost Climatic Stabilization (revised edition).* Rocky Mountain Institute, 1991. More recent studies by the Rocky Mountain Institute confirm the findings of the original study.

10 Pembina Institute for Appropriate Development. Sylvie Boustie, Marlo Reynolds, and Matthew Bramley. *How Ratifying the Kyoto Protocol Will Benefit Canada's Competitiveness.* Pembina Institute, 2002.

11   Kirsty Duncan. 2006. "Climate Change, Kyoto, and Economic Possibilities." Retrieved
     March 26, 2006 from frasermackenzie.com/KYOTO/Kyoto-Duncan.pdf.
12   Edmonton Journal. "Business as Usual Isn't Good Enough." Shell Canada spokesperson
     Jan Rowley. *Edmonton Journal*, December 16, 2002, pages C4-5.
13   Greenpeace Canada. Oct. 8, 2002. "Apocalyptic Campaign by Klein Based on Mistruths — Why
     is Klein Afraid to Debate?" Retrieved March 26, 2006 from action.web.ca/home/
     gpc/alerts.shtml?x=23587&AA_EX_Session=66660cafb4f96d8460d9ea147b587b98
14   Exxonsecrets. 2005. "Organizations in Exxon Secrets Database." Retrieved August 26, 2005
     from exxonsecrets.org/html/listorganizations.php.

Chapter 2: Conventions and Assumptions

15   US Department of Commerce. May 2001. "Profiles of General Demographic
     Characteristics: 2000 Census of Population and Housing." Retrieved Feb. 23, 2006 from
     census.gov/prod/cen2000/dp1/2kh00.pdf.
16   Fuel data from: U.S. Department of Energy, Energy Information Administration. 2004,
     Nov. 18. 2001 National Energy Survey "A Look at Residential Energy Consumption in 2001"
     (enduse_consump2001.pdf). Retrieved on Sep. 9, 2005 from:
     eia.doe.gov/emeu/recs/recs2001/detailcetbls.html. Calculations based on data from: World
     Business Council for Sustainable Development and World Resources Institute. Electricity
     Emissions (Electricity.xls). Retrieved Oct. 29, 2005 from:
     ghgprotocol.org/templates/GHG5/layout.asp?type=p&MenuId=OTAx. US Energy
     Information Administration. Voluntary Reporting of Greenhouse Gases Program Fuel and
     Energy Source Codes and Emission Coefficients. Retrieved Sept. 9, 2005 from
     eia.doe.gov/oiaf/1605/coefficients.html. US National Institute of Standards and Technology.
     2005. Guide for the Use of the International System of Units (SI): B. 9 Factors for Units
     Listed by Kind of Quantity or Field of Science. Retrieved June 3, 2005 from:
     physics.nist.gov/Pubs/SP811/appenB9.html; US Energy Information Administration. 2005.
     Voluntary Reporting of Greenhouse Gases Program Fuel and Energy Source Codes and
     Emission Coefficients. Retrieved Sept. 9, 2005 from eia.doe.gov/oiaf/1605/coefficients.html.
17   Coniff, Richard. "Counting Carbons: How Much Greenhouse Gas Does your Family
     Produce?" *Discover*. August 2005, pp. 54-61; Earthcare. 2005. EarthCARE Water Related
     FAQs. Retrieved Nov. 24, 2005 from earthcarecanada.com/FAQs/Water_
     FAQ.asp#w8; BUNDjugend. *Die Wette. Wie Jugendliche das Klima Retten*. Druckerei
     Pachnicke, 1998; Carbon Busters calculations.
18   *House*: University of Michigan Center for Sustainable Systems. Steven Blanchard and Peter
     Reppe. *Life Cycle Analysis of a Residential Home in Michigan*. M.Sc. Thesis, 1998; *Car*:
     Heather L. Maclean and Lester B. Lave. "A Life-Cycle Model of an Automobile."
     Environmental Policy Analysis 1988 3 (7), pp. 322A-330A. Summary at: Institute for Lifecycle
     Environmental Assessment. Sept. 12, 2003. "Automobiles: Manufacture vs. Use". Carnegie
     Mellon University, 1998. Retrieved June 19, 2006 from ilea.org/lcas/macleanlave1998.html;
     Refrigerator. University of Michigan Center for Sustainable Systems. Juhta Alan Horie. "Life
     Cycle Optimization of Household Refrigerator-Freezer Replacement (Report No. CSS04-
     13)". University of Michigan, 2004. Available for free download from:
     css.snre.umich.edu/css_doc/CSS04-13.pdf; *Photovoltaic (Solar) Panel*: US Department of
     Energy. January 2004. "What is the energy payback for PV?" Retrieved March 7, 2006 from
     nrel.gov/docs/fy04osti/35489.pdf; *Computer*. Dreier and Wagner 2000, quoted in Lawrence
     Berkeley National Laboratory. 2005. "Optimization of Product Life Cycles to Reduce
     Greenhouse Gas Emissions in California." CEC-500-2005-110-F. Available for free down-
     load from energy.ca.gov/2005publications/CEC-500-2005-110/CEC-500-2005-110-F.PDF.

Note: Studies conducted on the energy required to build and use a computer vary by as much as a factor of three. Energy costs of disposal and other secondary energy costs excluded from all graphs.

19 *Environmental Resources Management.* "Streamlined Life Cycle Assessment of Two Marks and Spencer plc Apparel Products." Marks & Spencer plc, January 2002.

Chapter 4: Your Family's Carbon Pie

20 See 16.

21 US Department of Energy. 2005. "Annual Energy Use in Residential Buildings." Retrieved Aug. 16, 2005 from: eere.energy.gov/buildings/info/homes/piehomes.html; Data are for 1997.

22 See 21.

23 Dieter Seifried. *Gute Argumente: Energie.* Beck, 1991.

24 See 17 "Counting Carbons".

25 See 17.

26 Based on a sedan driving 12,000 miles per year, and an SUV driving 15,000 miles in its first year, 12,000 miles per year thereafter. In 1994, an older person whose children left home drove on average only 8,600 miles, while a household with teenagers and an income of $50,000 or more drove 40,200 miles. See 17 "Counting Carbons"; US Energy Information Administration. Feb. 1, 2002. Chapter 3: "Vehicle-Miles Traveled." Retrieved June 28, 2006 from eia.doe.gov/emeu/rtecs/chapter3.html.

27 Energy Information Administration. 2005. November 2005 *Monthly Energy Review.* USEnergyPrices.pdf. Retrieved on Nov. 24, 2005 from eia.doe.gov/emeu/mer/prices.html; Energy Information Administration. 2005. November 2005 *Monthly Energy Review.* USEnergyPrices.pdf. Retrieved on Nov. 24, 2005 from eia.doe.gov/emeu/mer/prices.html; City of Albuquerque. 2005. *Water Systems: Water Trivia.* Retrieved on Nov. 24, 2005 from cabq.gov/water/trivia.html; Carbon Busters calculations.

28 Coldwell Banker Real Estate Corporation. 2005. Coldwell Banker(R) 2005 Home Price Comparison Index Reveals $1.7 Million Difference for the Same Size Home in Most Expensive and Most Affordable Markets. Retrieved on Nov. 25, 2005 from biz.yahoo.com/prnews/050922/nytho37.html?.v=23T; Carbon Busters Calculations.

Chapter 5: Home Improvement: Green Home Design

29 Rocky Mountain Institute. 2005. "RMI's Approach to Energy." Retrieved Nov. 28, 2005 from rmi.org/sitepages/pid116.php; Rocky Mountain Institute. "Tunneling Through the Cost Barrier". *Rocky Mountain Institute Newsletter.* Vol. 12 (2), Summer 1997, pp. 1-4; *Natural Capitalism: The Next Industrial Revolution.* Paul Hawken, Amory B. Lovins, L. Hunter Lovins. Earthscan Publications, 2000.

30 Lund University Institute of Technology. 2005. "Houses without Heating Systems: 20 Low-Energy Terrace Houses in Gothenburg, Sweden". Retrieved Nov. 8, 2005 from www2.ebd.lth.se/avd%20ebd/main/Gothenburg/Folder_Lindas_EN.pdf.

31 Paul Scheckel. *The Home Energy Diet: How to Save Money by Making your House Energy-Smart.* New Society Publishers, 2005.

32 Rocky Mountain Institute. *Visitors' Guide (3ʳᵈ Ed.).* Rocky Mountain Institute, 1991; Rocky Mountain Institute. *Visitors' Guide.* Rocky Mountain Institute, 2003.

33 Ted Rieger and Jeanne Byrne. "Off-Grid in a Cold City: The Alberta Sustainable Home." *Home Energy Magazine Online* March/April 1996; Jorg Ostrowski, pers. comm. May 16, 2005.

34 Rocky Mountain Institute. 2005. International Netherlands Group (ING) Bank, Amsterdam, Netherlands. Retrieved Dec. 2, 2005 from rmi.org/sitepages/pid208.php; Ernst Ulrich von

Weizsäcker, Amory B. Lovins and L. Hunter Lovins. *Faktor Vier: Doppelter Wohlstand, Halbierter Naturverbrauch.* Droemersche Verlagsanstalt, 1995; Amory Lovins. *Energy Efficiency.* Public lecture at Edmonton Space Science Centre, Oct. 30, 1995.

Chapter 6: Green Transportation

35 See 34, *Faktor Vier.*

36 CarSharing Network. 2005. "What is Car Sharing?" Retrieved Dec. 3, 2005 from carsharing.net/what.html.

37 Amory B. Lovins, E. Kyle Datta, Odd-Even Bustnes, Jonathan G. Koomey, and Nathan J. Glasgow. *Winning the Oil Endgame: Innovation for Profit, Jobs and Security.* Earthscan Publications, 2005. (Available for free download from: oilendgame.com/ ReadTheBook.html; see 34, *Faktor Vier.*

38 Amory B. Lovins and David R. Cramer. "Hypercars®, Hydrogen, and the Automotive Transition." *International Journal of Vehicle Design.* 35:1/2 (2004): 50-85. (Available for free download at rmi.org/sitepages/pid175.php#T0401.)

39 Ibid.

40 Ross Finlay. Sept. 18, 2002. "Eco-Speedster Stresses Both." Retrieved Oct. 29, 2002 from carkeys.co.uk/features/FE000450.htm; *Auto & Motor.* Sept. 28, 2002. "Opel Eco-Speedster: 250 km/h und 2,5 Liter Verbrauch: Opel Demonstriert mit Neuer Speedster-Studie sein Diesel-Know-How." Retrieved Oct. 29, 2002 from automotor.aol.de/automotor/auto/ specials/designstudien/index.jsp.

41 *Stern.* 2002. "VW Lüftet das Geheimnis um das 'Ein-Liter-Auto' ". Retrieved on April 15, 2002 from stern.de/sport-motor/auto/news/artikel/index.jsp?id=155293; Volkswagen. 2002. "Mobilität und Nachhaltigkeit: Die Zukunft ist Da — Volkswagen Baut das Erste 1-Liter-Auto der Welt." Retrieved on Dec. 14, 2005 from volkswagen-umwelt.de/buster/buster.asp?i=_ content/wissen_303.asp.

42 MSRP and fuel efficiencies retrieved Dec. 14, 2005 from edmunds.com. Fuel consumption figures used are EPA averages weighted 55 percent for city and 45 percent for highway driving, divided by 1.15 to simulate real-life fuel economy (The EPA is planning to update its rating system in 2006 to more closely approximate real-life consumption, which appears to be consistently higher than rated: US Environmental Protection Agency. Dec. 2005. *Green Vehicle Guide.* Retrieved Dec. 13, 2005 from epa.gov/greenvehicles/.

43 Natural Resources Canada. 2005. "EnerGuide Award Winners: The Most Fuel-Efficient Vehicles for Model Year 2006." Available for free download from oee.nrcan.gc.ca/ transportation/personal/choose_vehicle.cfm?attr=8.

44 See 42; Neil, Dan. Feb. 1, 2006. "On the Hybrid Horizon: Does it take AA or AAA batteries?" Retrieved Feb. 3, 2006 from latimes.com/news/printedition/highway1/ la-hy-battery1feb01,0,5437539.story?coll=la-news-highway_1.

45 *Incremental Hybrid Cost: Greenhybrid.* Dec. 22, 2005."Honda to Cut Civic Hybrid Costs by 1/3." Retrieved on Dec. 23, 2005 from greenhybrid.com; World Business Council for Sustainable Development. National Post. January 21, 2005. "Automakers Bank on Green." Retrieved Feb. 9, 2005 from wbcsd.org/includes/getTarget.asp?type=DocDet&id=12893. Now at wbcsd.org/plugins/workspace/message.asp?WspaceId=NjE&msgId=NDIyMA. *Hybrid Purchase Incentives:* Brendan I. Koerner, 2005. "Rise of the Green Machine." *Wired* magazine13.04, April 2005. Also available at newamerica.net/index.cfm?pg= article&DocID=2309.

46 honda.com; InsightCentral. 2003. "The Insight Story." Retrieved on Dec. 23, 2005 from insightcentral.net/; Edgar, Julian. May 19, 2003. Special Feature. "Honda Insight: An Incredibly Advanced Car." Retrieved Dec. 23, 2005 from autospeed.drive.com.au/cms/

A_1766/article.html. Wikipedia. March 20, 2006. "Toyota Prius." Retrieved March 20, 2006 from en.wikipedia.org/wiki/Toyota_Prius.

47   Colorado Department of Revenue, Taxpayer Service Division. Nov. 2005. "FYI Income: 9. Alternative Fuel Income Tax Credits." Retrieved Dec. 22, 2005 from revenue.state.co.us/ fyi/html/income09.html; Hyperion. 2005. "Hyperion Named Blue Sky Merit Winner for Its Drive Clean To Drive Change Program: CALSTART Award Recognizes Pioneering Initiative to Help Employees Purchase Fuel Efficient Cars." Retrieved Dec. 22, 2005 from hyperion.com/company/news/news_releases/press_release_2005_000659.cfm; Stuart Waterman. 2005. "Companies Help Employees Buy Fuel-Efficient Cars. Retrieved Dec. 22, 2005 from autoblog.com/2005/11/07/companies-help-employees-buy-fuel-efficient-cars.

48   Astoundingly, both *Consumer Reports* and automotive web site Edmunds.com initially released reports in 2005 claiming that hybrids deliver "little or no savings". However, *Consumer Reports* reversed its opinion in January of 2006, and Edmunds.com had based its analysis only on the first five years of owning the car, and assumed higher depreciation for hybrids than other cars (the hot market for used hybrids seems to indicate that the opposite is true). Sharp rises in gas prices since the release of these reports have tipped the balance even further in favor of the efficient hybrids described here.

49   US Department of Energy. November 2005. "State Financial Incentives and Laws: California." Retrieved March 20, 2006 from eere.energy.gov/cleancities/vbg/progs/laws2.cgi.

50   See 43.

51   Tom McCahill. "MI Tests the '54 Mercury." *Mechanix Illustrated*. March 1954 pp. 86-89.

52   Edmunds. 2006. "New Car Pricing: Hybrids: All." Retrieved March 22, 2006 from edmunds.com/apps/vdpcontainers/do/ViewMarketModels/category=market/attribute=hybrid.

53   Cutler J. Cleveland, Robert Costanza, Charles A.S. Hall and Robert Kaufmann. "Energy and the U.S. Economy: A Biophysical Perspective." *Science* 225 (4665): 890-897.

54   National Biodiesel Board. 2005. "Commonly Asked Questions." Retrieved October 21, 2005 from biodiesel.org/resources/fuelfactsheets/.

55   See 54.

56   National Biodiesel Board. 2005. "Biodiesel Myths and Facts." Retrieved October 21, 2005 from biodiesel.org/resources/fuelfactsheets.

57   Shell. 2004. "World-Beating Eco-Car Delivers 10,705 mpg: French Students Beat Own World Record at UK Eco-Marathon." Retrieved May 19, 2004 from shellglobalsolutions.com.

58   Lawrence Berkeley National Laboratory. Marc Ross and Tom Wenzel. March 2002. "An Analysis of Traffic Deaths by Vehicle Type and Model." Retrieved Dec. 2, 2005 from aceee.org/pubs/t021full.pdf; Gabriel Bridger. 2005. "Crash Testing: MINI Cooper vs. Ford F150." Retrieved Dec. 2, 2005 from bridger.us/2002/12/16/ CrashTestingMiniCooperVsFordF150.

Chapter 7: Electric Power

59   Energy Information Administration. 2005. November 2005. "Monthly Energy Review: USEnergyPrices.pdf." Retrieved on Nov. 24, 2005 from eia.doe.gov/emeu/mer/prices.html.

60   E Source. *E Source Technology Atlas Series Volume I: Lighting*. McGraw-Hill Companies, 1997; Philips. *Lamp Specification & Application Guide 2004*. Philips, 2004; TCP. *TCP 2005/2006 Product Catalog*. Technical Consumer Products 2005.

61   Warren E. Hathaway. *A Study Into the Effects of Types of Light on Children: A Case of Daylight Robbery*. Hathaway Planning & Consulting Services, 1994. Available for free download as Internal Report # 659 from the Canadian National Research Council at: irc.nrc-cnrc.gc.ca/pubs/fulltext/ir659/hathaway.pdf.

62   Ianto Evans. Nov. 8, 2004. pers. comm..

63 "Never say never." There probably are exceptions, but these rules should cover 99 percent of all equipment.

64 EPA. Nov. 23, 2005. "Refrigerators & Freezers: Energy Star." Retrieved Nov. 22, 2005 from energystar.gov/index.cfm?c=refrig.pr_refrigerators.

65 See 18, *Refrigerator*.

66 Rocky Mountain Institute. 2004. "Home Energy Briefs #6: Cleaning Appliances." Retrieved Feb. 15, 2006 from rmi.org/images/other/Energy/E04-16_HEB6CleaningApps.pdf.

67 See 32.

68 E Source. *E Source Technology Atlas Series Volume V: Residential Appliances*. McGraw-Hill Companies, 2001.

69 See 32, 68; Kuhn Rikon. 2006. Durotherm Thermal Cookware. Retrieved Feb. 22, 2006 from kuhnrikon.com/products/duro/group.php3?id=3.

70 See 68.

71 See 66; Michael Bluejay. Dec. 2005. "How To Save Money on the Use of Washers and Dryers." Retrieved Feb. 15, 2006 from michaelbluejay.com/electricity/laundry.html.

72 See 18, *Computer*.

73 Energy Star. 2006. "About ENERGY STAR." Retrieved Feb. 24, 2006 from energystar.gov/index.cfm?c=about.ab_index.

74 Apple Computer. Jan. 7, 2003. *12-inch PowerBook G4: Specification Sheet*. Retrieved Feb. 24, 2006 from download.info.apple.com/Apple_Support_Area/Manuals/specs/powerbook/L25519A_EN.pdf.

75 Acer. 2006. *TravelMate 3000: Technical Specifications*. Retrieved Feb. 24, 2006 from us.acer.com/acereuro/page9.do?dau34.oid=8961&UserCtxParam=0&GroupCtxParam=0&dctx1=25&CountryISOCtxParam=US&LanguageISOCtxParam=en&crc=2139838728.

76 Five-year savings: $343.18. Entry-level desktop system with 17-inch monitor, keyboard, and mouse, price after rebate: $299. Retrieved Feb. 23, 2006 from configure.us.dell.com/dellstore/config.aspx?c=us&cs=19&l=en&oc=DB110A&s=dhs.

77 Michéle Elsen 2006, pers. comm.

78 Shawn Murphy 2006, pers. comm.

79 US Department of Energy. Sept. 30, 2005. "Home Office and Home Electronics." Retrieved Feb. 25, 2006 from eere.energy.gov/consumer/tips/home_office.html; Lawrence Berkeley National Laboratory. 2006. "Home Office Equipment." Retrieved Feb. 25, 2006 from hes.lbl.gov/hes/makingithappen/no_regrets/homeoffice.html.

80 Jochen A. Siegle. Oct. 30, 2004. "Flüssiges Gold." Retrieved on Nov. 2, 2004 from spiegel.de; Dan Littman. Sept. 2003. "Cheap Ink Probed." Retrieved Feb. 23, 2006 from pcworld.com/news/article/0,aid,111767,00.asp.

81 IPKat. Sept. 1, 2003. "Rolling Back IP Rights?" Retrieved Feb. 24, 2006 from ipkitten.blogspot.com/2003_09_01_ipkitten_archive.html; Corante. Donna Wentworth. Sept. 2, 2005. "The Latest IP Crime: 'Box-Wrap' Patent Infringement." Retrieved Feb. 25, 2006 from copyfight.corante.com/archives/2005/09/02/the_latest_ip_crime_boxwrap_patent_infringement.php.

82 See 80, "Flüssiges Gold," and 81 "Rolling Back IP Rights?"

83 See 80, "Cheap Ink Probed."

84 Energy Star. 2006. *Printer & Fax Machine Key Product Criteria*. Retrieved Feb. 25, 2006 from energystar.gov/index.cfm?c=printers.pr_crit_printers.

85 Energy Star. 2006. *Printers*. Retrieved Feb. 25, 2006 from energystar.gov/index.cfm?fuseaction=find_a_product.showProductGroup&pgw_code=PR.

86 Guy Dauncey. *Going Carbon Neutral: A Guide for Publishers (Metric Edition)*. New Society Publishers, 2005. Retrieved March 7, 2006 from newsociety.com/Publishers%20CO2%20Template%20Metric.pdf.

87 Energy Star. 2006. *Copiers.* Retrieved Feb. 25, 2006 from energystar.gov/index.cfm?
    fuseaction=find_a_product.showProductGroup&pgw_code=CP.

88 Carbon Dioxide Reduction Edmonton (CO2RE). *CO₂RE Home$avers: Conserving
    Electricity.* Office of the Environment, City of Edmonton, undated.

89 Incremental costs per kWh for the nine top US suppliers ranged from 0.33 cents/kWh to
    1 cent/kWh. Incremental costs of five Canadian green power suppliers ranged from
    1 cent/kWh (Pembina Institute) to 4 cents/kWh (Ontario Green Tags). Averages quoted
    represent a simple average of offered prices on March 10, 2006, and are not weighted for
    volume of electricity delivered. Claudia Bolli 2006, pers. comm.

90 US Department of Energy, Energy Efficiency and Renewable Energy. 2005. "re_goals_work-
    shop_seattle_patton.pdf." Retrieved Nov. 12, 2005 from: eere.energy.gov/femp/pdfs/re_
    goals_workshop_seattle_patton.pdf.

91 See 18, "What is the Energy Payback for PV?"; see 53.

92 Natural Resources Canada. 2006. "Welcome to the Renewable Energy Deployment Initiative
    (REDI)." Retrieved Oct. 21, 2005 from www2.nrcan.gc.ca/es/erb/erb/english/View.asp?x=455,
    2005-10-21; Co-op America. March 2005. "Tax Credits for Cleaner Energy." Retrieved Oct. 21,
    2005 from coopamerica.org/pubs/realmoney/articles/greenenergyincentives.cfm, 2005-10-21.

93 Solar Energy Industries Association. 2006. "SEIA Releases Guide to New Federal Tax Credits
    for Solar Energy." Retrieved March 25, 2006 from seia.org/solarnews.php?id=96.

Chapter 8: Green Heating

94 Natural Resources Canada, Office of Energy Efficiency. *Air-leakage Control.* Natural
    Resources Canada, 1998.

95 American Council for an Energy-Efficient Economy (ACEEE). Alex Wilson, Jennifer Thorne,
    and John Morrill. *Consumer Guide to Home Energy Savings (8ᵗʰ Ed.)* ACEEE, 2003.

96 See 95.

97 Sun Power Consumer Association. 1989, quoted in: E Source. *E Source Technology Atlas
    Series Volume III: Heating.* McGraw-Hill Companies, 1996.

98 See 95.

99 California Energy Commission. 2006. *Geothermal Heat Pumps.* Retrieved March 28, 2006
    from consumerenergycenter.org/home/heating_cooling/geothermal.html.

100 Geothermal Heat Pump Consortium. 2006. "Commonly Asked Questions about an Uncommonly
    Sound Technology." Retrieved March 25, 2006 from geoexchange.org/about/questions.

101 Interstate Renewable Energy Council. Feb. 17, 2006. "Incentives for Renewable Energy."
    Retrieved March 7, 2006 from dsireusa.org/library/includes/incentive2.cfm?Incentive_
    Code=US37F&State=Federal&currentpageid=1.

102 Clark Public Utilities. February 2002. "Gas Pilot Lights." Retrieved March 28, 2006 from
    clarkpublicutilities.com/Residential/TheEnergyAdviser/Archives2002/2-02-1.

103 Office of Energy Conservation (Saskatchewan). 2006. "Hot Water Tanks." Retrieved March
    7, 2006 from oec.ca/html/programs/Housing/hot_water_tanks/index.cfm.

Chapter 11: Environmental Goods and Services

104 Canadian Parks and Wilderness Association. January 15, 2004. "Conservationnet: Grizzly
    Hunt, FSC, and More." Received January 15, 2004 from cpaws-edmonton.org.

105 William E. Rees May 11, 2005. "Urban Eco-Footprints and the Vulnerability/Sustainability
    of Cities: An Invitation to Change." Presentation to the Alberta Chapter of the Canadian
    Green Building Council, Edmonton, Alberta.

106 AAAS Atlas of Population and Environment. 2006. "Introduction American Association
    for the Advancement of Science." Retrieved March 7, 2006 from
    atlas.aaas.org/index.php?part=2.

107 WasteCap of Massachusetts. 2006. "Information on Recycling Paper." Retrieved on March 6, 2006. wastecap.org/wastecap/commodities/paper/paper.htm; and numbers from the United States Environmental Protection Agency.

Chapter 12: Putting It All Together

108 US Bureau of the Public Debt. 2006. "Recent Treasury Note and Bond Auction Results." Retrieved July 27, 2006 from wwws.publicdebt.treas.gov/AI/OFNtebnd; Millman. 2006. "2006 Pension Funding Study." Retrieved July 27, 2006 from milliman.com/pension_fund_survey/pfs_index.php.

Appendix F

109 See 17.

Appendix G

110 Energy Information Administration. 2005. "November 2005 Monthly Energy Review: USEnergyPrices.pdf." Retrieved on Nov. 24, 2005 from eia.doe.gov/emeu/mer/prices.html; Energy Information Administration. 2006. "Weekly U.S. Retail On-Highway Diesel Prices." Retrieved March 9, 2006 from tonto.eia.doe.gov/oog/info/wohdp/diesel_web_chart_us.gif; calculated based on figures in City of Albuquerque. 2005. "Water Systems: Water Trivia." Retrieved on Nov. 24, 2005 from cabq.gov/water/trivia.html.

Appendix H

111 US Energy Information Administration. "Voluntary Reporting of Greenhouse Gases Program Fuel and Energy Source Codes and Emission Coefficients." Retrieved Sept. 9, 2005 from eia.doe.gov/oiaf/1605/coefficients.html; Carbon Busters calculations based on US National Institute of Standards and Technology. 2005. *NIST Guide to SI units*. Retrieved June 3, 2005 from physics.nist.gov/Pubs/SP811/appenB9.html; Carbon Busters calculations based on NIS and US Energy Information Administration. "Voluntary Reporting of Greenhouse Gases Program Fuel and Energy Source Codes and Emission Coefficients." eia.doe.gov/oiaf/1605/coefficients.html; World Business Council for Sustainable Development and World Resources Institute. "Electricity Emissions (Electricity.xls)." Retrieved Oct. 29, 2005 from ghgprotocol.org/templates/GHG5/layout.asp?type=p&MenuId=OTAx; National Biodiesel Board. 2005. "Commonly Asked Questions." Retrieved Oct. 21, 2005 from biodiesel.org/resources/fuelfactsheets.

Appendix I

112 Energy Information Administration. March 2002. "Updated State-and Regional-level Greenhouse Gas Emission Factors for Electricity (March 2002)." Retrieved July 12, 2006 from eia.doe.gov/oiaf/1605/e-factor.html; GHG Registries. Aug. 2005. "Challenge Guide" (p. 26. Table 4). Retrieved July 12, 2006 from ghgregistries.ca/assets/pdf/Challenge_Guide_E.pdf.

Appendix J

113 See 111 *NIST Guide to SI units*; Carbon Busters calculations based on *NIST Guide to SI units*; see 31; National Energy Board of Canada. December 30, 2005. "Energy Conversion Tables." Retrieved March 23, 2006 from neb-one.gc.ca/Statistics/EnergyConversions_e.htm#NaturalGas2; after US Energy Information Administration. "Voluntary Reporting of Greenhouse Gases Program Fuel and Energy Source Codes and Emission Coefficients." Retrieved Sept. 9, 2005 from eia.doe.gov/oiaf/1605/coefficients.html.

Appendix K

114 See 111: *NIST Guide to SI units*.

Index

About the Author

GODO STOYKE is an award-winning environmental researcher and presenter with a Master of Science degree from the University of Alberta. Godo has been sharing his passion for sustainable design and education for the past 17 years. As president of Carbon Busters (carbonbusters.org), he has saved clients over 121 million pounds of CO_2 and $16 million in utility bills in North America and Europe. Godo has been living in an off-grid solar powered home for the last 16 years with his wife, Shanthu, and son, Calan, and has cut his heating and water consumption by over 50%, his gasoline consumption by 63% and his power consumption by 93%. He currently drives a gas-electric Prius and believes strongly in the role of the individual in shaping our common future.

Godo is a LEED (Leadership in Energy and Environmental Design) Accredited Professional registered with the US Green Building Council.

If you have enjoyed *The Carbon Buster's Home Energy Handbook*
you might also enjoy other

BOOKS TO BUILD A NEW SOCIETY

Our books provide positive solutions for people
who want to make a difference. We specialize in:

Environment and Justice • Conscientious Commerce • Sustainable Living
Ecological Design and Planning • Natural Building & Appropriate Technology
New Forestry • Educational and Parenting Resources • Nonviolence
Progressive Leadership • Resistance and Community

New Society Publishers

ENVIRONMENTAL BENEFITS STATEMENT

New Society Publishers has chosen to produce this book on Enviro 100, recycled
paper made with **100% post consumer waste**, processed chlorine free, and old
growth free.

For every 5,000 books printed, New Society saves the following resources:[1]

19	Trees
1,703	Pounds of Solid Waste
1,874	Gallons of Water
2,444	Kilowatt Hours of Electricity
3,095	Pounds of Greenhouse Gases
13	Pounds of HAPs, VOCs, and AOX Combined
5	Cubic Yards of Landfill Space

[1]Environmental benefits are calculated based on research done by the Environmental Defense Fund and
other members of the Paper Task Force who study the environmental impacts of the paper industry.

For more information on this environmental benefits statement, or to inquire about environmentally
friendly papers, please contact New Leaf Paper – info@newleafpaper.com Tel: 888 • 989 • 5323.

For a full list of NSP's titles, please call **1-800-567-6772** *or check out our website at:*

www.newsociety.com

NEW SOCIETY PUBLISHERS